THE
SECRET
HISTORY
OF THE
TWENTIETH
CENTURY

HERBERT G. DORSEY III

Table of Contents

Prologue

Most people might wonder why one would bother to write a secret history of the Twentieth Century. What could be secret about a well recorded history? Isn't this what we are taught in history courses in school and in books on history?

Well unfortunately, this sort of history usually states part of what happened, while giving names, dates and so on. But for the large part, the causes of these happenings are glossed over or not even discussed at all. Often the causes behind the history are hidden by secret societies to hide their own role in what often amounts to unjust, immoral or criminal activities.

At no time is this truer than in the Twentieth Century, where because of so called "national security" much has been hidden from the public. This secrecy became dramatically pronounced in the second half of the Twentieth Century following the great changes caused by World War II and keeps increasing to the present time.

Great scientific and technical advancement also occurred during the Twentieth Century, as most of us realize. However, more of this advancement has been hidden than revealed to the public and this advancement is much greater than most even realize.

And of course, much of this hiding of history, science and technology is being done by the leaders of large multinational corporations which now have more actual power than the nation states which incorporated them. The largest corporation on the planet, which also is a sovereign country and poses as a religion is the Vatican. And as we shall see, the Vatican is controlled by another force entirely.

Before writing about the secret history of the Twentieth Century, some precursor information is required to fully understand the causes of the events, many of which are still hidden, of that eventful century. So, I have included two Chapters; "Prehistory" and "Before the Twentieth Century", to bring readers up to speed.

"Prehistory" is a rough outline covering millions of years and was obtained from several esoteric sources that explain the extraterrestrial source of ancient civilizations on our planet. To some this will sound like science fiction. I will include evidence that it is not!

"Before the Twentieth Century" covers the evolution of hidden political forces that largely affect our world in the present. Particularly, I cover the Illuminati – probably the largest secret political and economic force on the planet.

I do not intend to cover all the history of the Twentieth Century in this treatise. That would be too large a project. I will primarily focus the more hidden aspects of this history, which I think will prove interesting and give the reader a more comprehensive understanding of the causes behind this history.

My purpose in writing this book is to reveal that which has been hidden so that the public will become more knowledgeable of the world around them and thereby gain greater control over their own lives, fortunes and destiny. Read and learn!

Prehistory

According to many esoteric sources, humankind did not evolve on planet Earth. It evolved in other star systems and was then later transplanted to our planet. As revealed in Lyssa Royal's channeled book, *The Prism of Lyra* and in Stewart A. Swerdlow's book, *True World History: Humanity's Saga,* and in Len Kasten's *Alien World Order,* humankind evolved in the Lyran star system, the brightest star of which is Vega.

Another group which had a large effect in our planet from prehistoric times to the present are the Draco Reptilians, which evolved in the Draco constellation of stars, primarily Alpha Draconis. In appearance, they are large, about 8 feet, and muscular and have a scaly reptilian skin with vertical slit eyes. The space faring Humans from Lyra and the Draco Reptilians have been in constant conflict for millions of years.

The Dracos first attacked the peaceful humans of Lyra causing the Lyrans to scatter and set up refugee colonies on planets of other star systems.

And, both are presently dwelling on our planet – Humans on the surface of our planet and the Reptilians hidden under the surface in large, deep cave systems containing entire cities.

The Reptilians by no means are the only extraterrestrials and intraterrestrials interacting with the Earth. More commonly seen E.T.s are the Greys, The Tall Whites, the Mantids and the human looking Nordics. People experiencing abduction experiences commonly see the Greys and occasionally the Mantids. While certain military personnel interact with the Tall Whites who seem to have a hidden colony near the Indian Springs, Nevada military reservation and with the Reptilians in their underground bases – some jointly operated by the Reptilians and the U.S. military.

Also, there are other varieties of E.T.s. According to Clifford Stone, who worked on highly classified crashed UFO recovery operations for the military, the U.S. military has catalogued 57 different types of E.T.s

However, it is the Reptilians that have primarily impacted the history of our planet. Also, some of the Greys seem to be under the influence of the Reptilians. So, the Reptilians will be the focus of this history,

The Reptilians had been making war on the Lyran star system for a long time, which forced many Lyran refugees to colonize other planets in other star systems. The Lyran settled two planets in our solar system which were Mars and the now extinct Maldek.

Mars was once a moon of Maldek with plenty of water and atmosphere. The humans which settled and evolved there were smaller in size because of the lighter gravity. On Maldek, the humans evolved into a more giant size because the planet was much larger and they evolved in a higher gravity.

Maldek was destroyed in one of the wars between the Reptilians and the Humans and the debris of that planet now make up the asteroid belt between Mars and Jupiter. Mars no longer orbited Maldek and assumed it's present orbit around the Sun. The Reptilian weapon which destroyed Maldek was a planet sized space ship disguised as an ice comet – a veritable death star. Some of the people of Maldek saw through the disguise and knew what was coming and took refuge on Mars before the destruction of their planet.

Most of the atmosphere of Mars and much of the water was blown away by this planet's destruction and the side of Mars facing Maldek was pock marked with many craters. Humans did survive this catastrophe on Mars by moving underground for protection from the increased radiation due to a thinner atmosphere. Now, there were both Martians and the larger Maldekians living underground on Mars.

NASA had a computer program project called *Brilliant Pebbles,* where the trajectories of all the asteroids in the asteroid belt were placed in the program and their trajectories computed back in history to a distant time when all the asteroids were in one location. That location was the former planet Maldek.

On Earth, the continents of Atlantis and Lemuria both emerged from beneath the oceans about one million years ago. The Draco Reptilians started settling in Lemuria after flora and fauna became established there, less than a thousand years after the continent surfaced above the Pacific Ocean. These Reptilians had brought another of their planet sized weapon – the Moon – into Earth orbit and settled Lemuria from their huge space ship. One of the weapons of choice of the Reptilian alliance are hollowed out planets and comets which sometimes allow a stealthy approach.

The Reptilians settled our planet first on the now submerged continent known as Lemuria. For this reason, the Reptilians claim the Earth is theirs. They basically had the whole planet and it's "moon" to their selves. However, the Reptilians, for whatever reason, pretty much remained on the Moon, Lemuria and nearby regions.

Long after the Reptilians settled in Lemuria, about three hundred thousand years ago, the Lyrans sent a particularly troublesome and warlike group of humans, known as the Atlans from the Pleiades, to planet Earth with the specific intent of confronting the Reptilians. The Atlans settled in the continent named after them, called Atlantis, which is now also submerged beneath the Atlantic Ocean.

By this time, the Lyrans had formed 110 refugee colonies on other star systems and an anti- Draco Reptilian alliance called the Galactic Federation of Light. The troublesome and independent Atlans refused to join this Federation, so the Lyrans decided to rid themselves of these troublesome people and put them to good use confronting the Reptilians on Earth. The Atlans were unafraid of the Reptilians and possessed advanced war making technology.

The Federation's plans soon materialized as open warfare between Atlantis and Lemuria on planet Earth. This war went on for many thousands of years, with the Atlans developing ever more powerful weapons during the conflict. Eventually, about fifty thousand years ago, they developed a powerful electromagnetic pulse weapon so powerful that it caused the eventual submergence of the entire continent of Lemuria.

Reptilian survivors moved to nearby areas like Australia, the Philippines, the Tibetan sub-continent and Antarctica. They moved to and created underground cities there to hide from the victorious Atlans.

After this huge victory, Atlantis flourished. The Atlans also colonized parts of Egypt and Central America. This golden age of Atlantis continued until a series of geophysical disasters struck Atlantis starting about fifteen thousand years ago and culminating about eleven and a half thousand years ago, when Atlantis finally completely submerged beneath the Atlantic Ocean.

Some of this disaster was caused by similar weapons used by the Atlans to sink Lemuria, which were used against the underground Reptilian bases, which had the unintended effect to also disrupted the geology beneath Atlantis.

During this golden age of Atlantis, Martians and Maldek survivors also came to Earth. This influx from Mars caused a small race of giant humans on the planet. Their skeleton remains are occasionally discovered and quickly snapped up by the Smithsonian Institute and hidden away from public view. Apparently, for whatever reason, this institute wishes to hide evidence of the real history of our planet from the public.

The Martians went to Sumer and created a full-blown civilization there - just as the Atlans were creating a full-blown civilization in Egypt as a colony.

The Maldekian survivors settled in what now is the Gobi dessert. Now, they also were making war on the Reptilians who remained on the Moon. Apparently, the Reptilians possessed nuclear weapons in their hollowed-out Moon base and launched them against the Maldekians in the Gobi dessert. In places, the sand was turned into glass – much like at the Nevada atomic test site.

This warfare forced the Maldekians to also go to and build many underground cities. One of their cities is under the Tibetan mountains and another is under mount Shasta, where they still live as a race now called the Agarthans. They prefer to remain underground because of mankind's war like activities on the surface of this planet.

Eventually the Reptilians on the moon were defeated by the Maldekian forces allied with the Glactic federation of light. However, a later Galactic treaty agreement turned the Moon into a diplomatic meeting area. This was because our Sun is also a traversable wormhole portal, allowing rapid interstellar travel within our galaxy. So, our solar system has a higher than usual number of interstellar travelers.

This made the, then vacant, moon an excellent meeting place to carry out diplomatic issues. Even if different star systems are at war elsewhere, they can have bases on the Moon and exist peacefully, within kilometers of each other. That is how strictly the Moon diplomatic zone is administered. The Reptilians are still allowed a sector of the Moon and have bases there for these diplomatic processes. This is based on information from Corey Goode and others that have been to the Moon at the Lunar Operations Command.

In their underground city of capital city of Bhogovita, which connects to other underground cities via tunnel systems, beneath India, the Reptilians had developed mind control techniques known as telepathic hypnosis which could be used to influence unsuspecting humans on the surface.

From their underground bases in India, they used this telepathic hypnosis to create the religious cult known as the Nagas, representing the snake people. Soon, the Atlans realized what was going on and were at war with the Nagas in India.

So, not all was peaceful on Earth after the sinking of Lemuria. In India, whole groups and their leaders were being mentally attacked by the Reptilians. Wars started between these mind controlled humans and the Atlans. These wars even included atomic weapons. In India, the records of these horrible atomic wars were recorded in the book of Dzyan and the Rig Veda.

Later, from their underground bases in other parts of the planet, the Reptilians would use their telepathic hypnosis to create other religious cults among the other surface populations. These religious cults invariably involved human sacrifice and often included snake and reptilian deities. The practices of the Aztecs of Central America would be a well-known example of these religious cults.

The concepts of money and centralized banking to control humankind is also a Reptilian inspired concept which engenders the emotion of greed and desire for money power over others among certain humans.

Most people can intuitively pick up the "vibes" that other people put out. These vibes are transmitted via the Astral plane. The Reptilians are even more attuned to these vibes. The Reptilians are quite cognizant of the Astral plane, where human emotions predominate. The baser human emotions such as anger, fear, sexual arousal, and greed, create energy fields on the lower Astral plane, which the Reptilians call "Loosh".

The Reptilian subtle body feeds off this Loosh. So, by inspiring wars, human sacrifice, pornography, and instilling fear, the Reptilians generate plenty of Loosh from humans, which energizes their own beings.

The more refined human emotions such as bravery, honesty, love and respect, loyalty and spirituality are repellent to these Reptilians, which have little or no understanding of these, spiritually higher, human traits. These higher traits also generate energy on the Astral plane of a higher and finer vibration – much too high for the reptilians to want or use!

The type of Astral energy habitually surrounding a human also determines where it's soul will exist after transitioning from the physical form, in the process called death. The Astral plane has several different gradations, from the very course, anguishing and confusing to the most serene and sublime. According to Robert A. Monroe, who has experienced many astral trips out of his body, and as described in his book, *Journeys Out of the Body*, in the Astral, and even more so in the Mental planes, you literally create your own heaven or hell by your own thoughts and emotions. Luckily, most, but not all, people are spiritually evolved sufficiently to avoid the lower Astral realms where disembodied Reptilians souls dwell.

The Reptilians also had an influence in later civilizations, such as Rome. The arenas, where gladiators fought to the death and Christians were torn apart by lions, were inspired by the Reptilians. The Roman punishment, by crucifixion, of human beings was also inspired by the Reptilians.

Later, when Christianity was made the legal religion of Rome, the non-Catholic versions of Christianity was called heresy and their worshippers were often killed in the cruelest ways, like burning at the stake.

When the Knights Templars were outlawed by the Pope by the urgings of King Phillip in France, they were tortured in the most horrible ways imaginable to extract confessions. Then, after confessing, they were burned at the stake. Some confessed to nothing and were burned anyway.

Similar Catholic practices continued during the Inquisition. Persons charged with heresy had all their property confiscated by the Church and then they were burned at the stake.

These practices were perversions of Christ's actual teachings and were inspired by the Reptilians using their telepathic hyp-nosis from their underground bases - one of which lays deep beneath the Vatican in Italy.

Before the Twentieth Century

To better understand what happened during the Twentieth Century, a quick look at important events which occurred before then is required. Many of these events I have written about in my book, "The Secret History of the New World Order". For those of you who haven't read that book, a quick review is required. I will merely touch on the high points here.

After the collapse of the Roman Empire, their legal religion of Christianity, or the Catholic version of it, continued on. By the time of Charlemagne, The Catholic Church became the ruling power of much of Europe, which was called the Holy Roman Empire. This rule was executed by the power of belief. The Church had no army of its own. But the kings of this empire actually believed that the Pope was Christ's representative on Earth and that the Pope literally held the keys to Heaven and Hell. Therefore, these kings were quite willing to abide by the Pope's will.

The Holy Roman Empire reached its pinnacle by the 1500s. But, a new religion, Protestantism was starting to spread in countries like England, Germany, Switzerland and France. And this resulted in loss of influence of the Catholic Church.

The Society of Jesus was commissioned by Alessandro Farnese or Pope Paul III on September 27, 1540, in response to the Protestant Reformation. The leader of this new society was Ignatius of Loyola, son of a Jewish family that had converted to the Catholic religion to prevent being expelled from Spain and losing the family fortune.

Ignatius of Loyola was himself a member of the Alumbrados, a secret society whose name meant the enlightened ones. And, he incorporated some of the secret methods of the Alumbrados into the formation of the Society of Jesus.

The Catholic Church was losing much influence in Europe due to the Protestants, who under Martin Luther and others, first tried to reform the church and then, failing in this, started their own versions of Christianity. Since these new forms of Christianity failed to recognize the primacy of the Pope, the Catholic Church considered them heretics.

The mission of the Society of Jesus was to combat heresy and restore the Catholic Church to its former glory and power in Europe. The Society of Jesus, often called the Jesuits, were so effective in their mission that, before long, they were placed in charge of the Inquisition, which was formerly headed by the Dominicans.

The Jesuits were organized as a strict military organization with strict obedience to their superiors. Not only the will of the Jesuit, but also reasoning and moral scruple must be sacrificed to this obedience. This obedience went up the chain of command right up to the Jesuit Superior General, who controlled the entire Society of Jesus worldwide. Soon, the Jesuits were granted powers that no other Catholic sect had. Pope Gregory VIII, in 1572, granted the Jesuits the power to do commerce. This was the only Catholic sect allowed that privilege since the Knights Templars were granted that privilege, four centuries earlier.

The fifth Jesuit Superior General, Claudio Acquaviva in 1580, ordered Jesuit, Father Vilela to purchase of the port of Nagasaki from a local Japanese warlord. Acquaviva then sent Jesuit, Alessandro Valignano back to manage the new commercial mission. The Jesuits promoted heavily the growth of their wholly owned port of Nagasaki, to one of the most profitable trading ports in the world. Jesuit ownership of the port of Nagasaki gave the Society a concrete monopoly in taxation over all imported goods coming into Japan.

Claudio Acquaviva formed an alliance in 1595 with the Dutch in supporting their merchant ships and trade. This resulted in the, Jesuit controlled, Dutch East India Company. Using the exclusive powers of the Jesuits to conduct banking and commerce, the Dutch East India Company represented one of the most profitable companies of history, thanks to its control of spices, drugs and plantations.

Claudio Acquaviva died at Rome in 1615, leaving the Society of Jesus nearly tripled in size and numbering 13,000 members in 550 houses and 15 provinces. The subsequent influence exercised by the Jesuits, in their golden age, was largely due to the far-seeing policy of Acquaviva, who is undoubtedly one of the greatest Superior Generals to have governed the Society.

One large competitor to the Dutch East India Company was the British East India Company. In the year 1600, Queen Elizabeth I presented a charter to the East India Company. Officially, its business was trading in tea. While this is certainly true, there is a much darker side to the story. The true goals of this company were two-fold: To study the banking and financial systems of the

Eastern world, (namely India) for eventual manipulation, and to introduce opium to the Far East. It is through the latter of these two that the East India Company was able to gain massive amounts of wealth.

The British began growing vast plantations of opium-producing poppies in her colony in Bengal, India. From there, the poppies were transported in the form of raw opium, via English Tea Clipper ships, to China, where opium was sold to the Chinese people. For many decades, this opium trade continued uninterrupted, and two main things occurred because of it: 1. The East India Company made a fortune selling these vast amounts of opium and 2. The Chinese nation became a nation of addicts, with many millions of people becoming addicted to opium and thereby halting progress to the build-up of the infrastructure of China. The government of China would later resist the opium trade, resulting in the Opium Wars.

The ruling body of the East India Company was known as The Council of 300, yet as the opium trade began to grow and become more lucrative, pulling in massive amounts of wealth, the British monarchy merged with the company, and from this emerged a group that to this very day has referred to themselves as The Committee of 300. The British "Secrecy Acts" were enacted to protect the monarchy from the scandal of the Royals being involved in opium trading.

Dr. John Coleman, an alleged former MI6 (British Military Intelligence Division 6) foreign secret service agent with top level security clearance to the upper echelon of the workings of the British aristocracy, has made it his life's work to 'rip the mask off' the hidden agendas of this organization.

Doctor Coleman is a best-selling author who has written many books and monograms on the subject, the magnum opus of which is entitled *Conspirator's Hierarchy: The Committee of 300*. I highly recommend that anyone wishing to know what is really going on in the world to read this book.

The Jesuits also created Jesuit universities to influence the thinking of young people, many of whom would leave the Protestant Church and rejoined the Catholic Church. They also used these universities to rewrite some of the history to hide what they had been doing.

The Pope had also given the Jesuits the office of the confessional and the power to absolve confessed sin. Confessions were supposed to be confidential between confessor and priest. However, if the confessor was a person of influence like a king, prince or baron, a record of the confession was kept and sent to the Jesuit Superior General at the Vatican. Often these records could be used as blackmail to further the Jesuit agenda. Since the Jesuits were in many countries, this practice became the first international intelligence agency.

On the even darker side, in countries like France and Ireland, which had both Protestants and Catholics, Jesuit priests would preach from the pulpit that it was no sin to kill a Protestant because they were heretics.

In some cases, these Jesuit priests would urge their parishioners to go out and slay heretics. This lead to large massacres of Protestants in Ireland and the massacre of the Hugonauts in France. This was a primary reason why many Protestants came to the American Colonies – to escape the Jesuit repression in Europe.

The Jesuits also were involved in considerable financial fraud, and political subversion. The infamous Gunpowder Plot was organized by the Jesuits to blow up the Protestant British Parliament and King for example.

This subversive activity on the part of the Jesuits caused them to be expelled from some European countries. And the kings of Europe complained to the Pope about their many crimes and demanded that the order be outlawed or else they would do like England and start their own Church. Finally, the Society of Jesus was outlawed in 1773, by order of the Pope.

However, an organization as wealthy and powerful as the Jesuits would not easily disappear.

In non-Catholic countries, the Jesuits merely continued as usual. In Catholic controlled countries, the Jesuits "went underground." By May 1, 1776, they had organized the Illuminati, with the Jesuit Adam Weishaupt as their leader. Recognizing the secret power of the Freemasons, they were also infiltrating Freemason lodges and gaining control over them.

Meyer Amshel Rothschild, a successful banker and Freemason, was persuaded to join the Illuminati. Later, agents of the Illuminati persuaded the Pope to appoint Rothschild to become the Guardian of the Vatican Treasury. This allowed the outlawed Jesuits to have access to the great wealth of the Vatican to finance their various projects.

These Jesuit controlled Freemason lodges then organized the American revolution and the French revolution as discussed in books like *Rulers of Evil* by Tupper Saussy and *The Secret History of the New World Order* by myself. Both Washington and Jefferson were ardent supporters of Adam Weishaupt, the Illuminati leader.

Before the American revolution, the British colonies were all Protestant, except Maryland, with Pennsylvania having a small minority of Catholics. Most of the Colonies prohibited Catholics from holding political office because these Protestants or, their forefathers, had fled Catholic prosecution in Europe. After the Revolution, the separation of Church and State clause of the Constitution allowed anyone to hold political office regardless of their religion. Thus, the revolution allowed the Jesuits and other Catholics to obtain more political power in the post-revolutionary United States.

The outlawed Jesuits also finally took over their competing British East India Company, through the Jesuit Lord Shelburne in 1783. Shelburne used some of these profits to finance another Jesuit project – the French Revolution.

The French Revolution was followed by a reign of terror with many losing their heads to the Guillotine. It was also followed by a strong anti-Catholic movement. Church Property was confiscated and clergy imprisoned. This was because the Jesuits, who were secretly leading the revolution, still had not forgiven the Catholic Church for outlawing them in 1773.

Also in France, the outlawed Jesuits were promoting a talented young military man, known as Napoleon Bonaparte. Later, they would command where his military campaigns would be directed. Three of these campaigns would directly benefit the outlawed Jesuits – the invasion of Austria, the Island of Malta and Italy.

The invasion of Austria, ended the Holy Roman Empire when the last Holy Roman Emperor, Francis II was defeated by Napoleon in the battle of Austerlitz in 1806.

The Vatican's military, the Knights of Malta, had their headquarters on the island of Malta. Their defeat by Napoleon at Malta essentially placed the Knights of Malta under Jesuit control.

After the Invasion of Italy, Napoleon had the Pope arrested and thrown into prison. The Jesuits understood the value of working under the cover of the Catholic Church, which had tremendous influence in the world. For this reason, they desired to have their order reinstated.

So, they had Napoleon inform the imprisoned Pope that if he would reinstate the Society of Jesus, he would regain his freedom and regal position as Pope. After 5 years in prison, the Pope finally relented and the Society of Jesus was reinstated in 1814. Since then, the Jesuit Superior General was the secret but real power at the Vatican, while the Pope was reduced to a mere figurehead.

After this, the Jesuits no longer needed Napoleon. Using their secret influence, the Jesuits arraigned for Napoleon's arrest and exile. After being betrayed by his Jesuit masters and imprisoned in St. Helena Island, in the middle of the South Atlantic, Napoleon had this to say:

> "The Jesuits are a military organization - not a religious order. Their chief is a general of an army, not the mere father abbot of a monastery. And the aim of this organization is power. Power in its most despotic exercise. Absolute power, universal power, power to control the world by the volition of a single man..."

The Illuminati also created other secret societies, like the Carbonari in Italy and Skull and Bones in Germany. In 1833, Chapter 322 of the Order of Skull and Bones was created at Yale

University in the United States, by two students returning from Germany that had been initiated into the Illuminati, William Huntington Russell, and Alfonso Taft. This chapter of Skull and Bones was incorporated as the Russel Trust in 1856, financed by Russel and Company, a large opium trading syndicate.

In Anthony Sutton's book, *America's Secret Establishment,* it is pointed out that a number of members of Skull and Bones were leading secessionist movements in the Southern States, leading to the Civil War. This was part of a larger Jesuit plan to destroy the United States, which had grown too independent for their liking. In my book, *The Secret History of the New World Order,* I document the number of U.S Presidents and other important U.S. political figures that were assassinated by the Jesuits.

After the Death of Adam Weishaupt in 1830, control of the Illuminati went to Giuseppe Mazzini in 1834. Mazzini was also head of the Carbonari, a secret resistance group to the tyranny imposed on France and Italy by the forces that defeated Napoleon at Waterloo.

This group was behind the Risorgimento movement that is credited with forming the Kingdom of Italy in 1861. In 1860, Mazzini also organized the Mafia.

At this time, the American Illuminati, under Scottish Rite Freemasonry, was headed by Albert Pike. On August 15, 1871, Albert Pike had written a letter to Mazzini which outlined his Illuminati plan to rule the world through the use of three world wars.

The First World War would pit the Germans against the British and overthrow the Czar of Russia and use communism to create a Godless country there.

The Second World War would pit Zionism against Nazism and create the State of Israel. Also during this war, International communism would be strengthened so that it was in strength equal to united Christendom.

The Third World would stir up the differences between political Zionism and the Muslims. This war was to be directed so that Zionism and Islam would destroy each other. (1)

This letter, written shortly after the Civil War by Albert Pike, showed surprising prescience in predicting events of the next century and demonstrates the amount of power and long range planning the Jesuit's Illuminati had in shaping world events.

The Beginning of the Twentieth Century

In the summer of 1905, President Theodore Roosevelt dispatched the largest diplomatic delegation to Asia in U.S. history. Roosevelt sent his secretary of war, seven senators, twenty-three congressmen, various military and civilian officials, and his daughter on an ocean liner from San Francisco to Hawaii, Japan, the Philippines, China, Korea, then back to San Francisco.

One notable person in this delegation was the U.S. Secretary of War, William Taft. William Taft's father, Alphonso Taft, was one of the persons to set up the Order of Skull and Bones at Yale in 1832. William Taft was also the past Governor of the Philippines, which the U.S. had taken from Spain in the Spanish American War.

Over the course of this imperial cruise, Theodore Roosevelt made important decisions that would affect America's involvement in Asia for generations to come. Theodore Roosevelt had been enthusiastic about American expansion in Asia, declaring, "Our future history will be more determined by our position on the Pacific facing China than by our position on the Atlantic facing Europe."

He was confident that American power would spread across Asia just as it had on the North American continent. Hawaii, annexed by the United States in 1898, had been the first step in that plan, and the Philippines was considered to be the launching pad to China. In 1898 during the Spanish American war, Filipino freedom fighters had expected that America would come to their aid in their patriotic revolution against their Spanish colonial masters.

Instead, the Americans short-circuited the revolution and took the country for themselves. And the U.S. ruled it even more cruelly than Spain, decimating whole villages that resisted the Yanks.

In 1901, on the Island of Samar in the village of Balangiga, 51 Americans had been killed. General Jake Smith, who was put in charge of disciplining Balangiga and the island of Samar, ordered all people over 10 years of age be killed on the whole island.

Afterwards, William Taft would claim that the insurrection had been put down. But, the Philippine freedom fighters fought a 4-decades long and bloody war for independence from the U.S. that lasted until just before World War II.

Taft—at Roosevelt's direction—brokered a confidential pact allowing Japan to expand into Korea. It is unconstitutional for an American president to make a secret treaty with another nation without United States Senate approval. But Roosevelt felt he could get away with it since Japan was so far away and all agreements with Japan were kept secret.

Theodor Roosevelt had already decided that Japan would be a good ally in the U.S. conquest of Asia. Like Hawaii, Japan would provide a coaling station to replenish steam ships on their long voyages across the Pacific.

Japan had learned to accept the need to modernize their military and naval power. Japan could no longer live in peaceful isolation after being given a taste of Admiral Perry's gunboat diplomacy which was used to open Japan to U.S. markets.

Japan had learned to imitate the ways of the U.S. after Perry had demonstrated the power of iron steamships with powerful cannons and was quickly developing their own modern military power while the rest of the Orient retained their ancient ways.

When Japan invaded Korea, the U.S. closed their Korean Embassy - while embassy officials and personnel left Korea like rats leaving a sinking ship.

Later, when Korea asked for help from the ally they trusted, the United States, they were informed that Korea was now officially part of Japan and all requests had to be routed through the Japanese government.

During the war between Japan and Russia, Theodore Roosevelt developed good communications with Japanese Emperor Meiji, who oversaw Japan's tilt toward the west and away from Asia. President Roosevelt sent him the skin of a Colorado bear he had killed while hunting as a token of Japan having good luck in their campaign against the Russian bear. It seemed to work. In the following naval battel, Japan sunk all the Russian ships without losing any of their own.

And as he was negotiating secretly with the Japanese, Roosevelt was simultaneously serving as the "honest broker" in discussions between Russia and Japan, who were then fighting what was up to that time history's largest war. The combatants would sign the Portsmouth Peace Treaty in that summer of 1905, and one year later, the president would become the first American to be awarded the Nobel Peace Prize.

Secretary Taft and Japanese Prime Minister Katsura met secretly in a simple, unadorned room in Shiba Palace. Taft, speaking for Roosevelt, asked that Japan would not bother the Philippines. Katsura assured Taft that Japan had no interest in the Philippines.

Taft then persuaded Katsura that Japan needed a doctrine like the Monroe Doctrine that would make Japan the protector of other Asian states from foreign interference and that this understanding would be backed up by both the U.S. and Briton.

The Japanese Prime Minister, responded that it would be good if such an understanding was backed up with a formal treaty between the three powers. Taft knew that the U.S. Senate would never ratify such a treaty.

Taft quickly added, "Without any agreement at all… just as confidently as if a treaty had been signed… appropriate action by the United States could be counted upon" to support Japan's sphere of influence in Asia because "the people of the United States were so fully in accord with the policy of Japan and Great Britain."

This remarkable American commitment to Japan's expansion "as if a treaty had been signed" would remain secret for almost two decades and has been obscured by time. In the long run, these events, caused by Theodor Roosevelt's penchant for U.S. imperialism, would set the stage for the Japanese attack on Pearl Harbor and the Korean War.

William Taft would later become the 27 President of the United States and a Supreme Court Justice. Many Skull and Bones members of Yale go on to important positions. Both George W. Bush and John Kerry are members of the Illuminati Society of Skull and Bones.

Another Illuminati organization was the British Roundtable, which was organized by Cecil Rhoads. Many Jesuits and Freemason students from Oxford University would be in this organization. Rhoads was himself a graduate of Oxford and was initiated into the Illuminati's Scottish Rite Freemasonry there.

Rhoads had eight secret wills. But one well known one, created the Rhoads Scholarship to Oxford. In a letter accompanying his fourth will, written in June 1888, Rhodes instructed Lord Nathaniel M. Rothschild, his collaborator and financier at De Beers and to whom he originally left most of his fortune - to obtain the Constitution of the Jesuits and "insert English Empire for Roman Catholic Religion" each place that it occurred. Then, the secret Roundtable society could use the modified Jesuit document as its own charter.

After the death of Rhoads, Nathanial Rothschild ran the Roundtable, as per Rhoad's fourth will. The Queen of England was a major supporter of this secret Roundtable. Nathanial Rothschild choose Lord Milner to help manage the Roundtable.

The stated goals of this Roundtable were to extend British rule over the world and peacefully return the American colonies back to British control. The Roundtable organized the Royal Institute of International Affairs (RIIA) in England and the Council on Foreign Relations (CFR) in the United States. Later creations of this Roundtable would include the Cub of Rome and the Bilderberg Group.

Also, the Federal Reserve Act of 1913 was planned by Nathaniel Rothschild at these secret Roundtable meetings. Later, the Progressive Income Tax for the U.S. would also be planned by this Roundtable.

One notable agent of Nathanial Rothschild was Edward Mandell House, who also was a member of the Roundtable. House would turn up in many interesting places and be involved in a number of history changing events. He was born in Houston, Texas on July 26, 1858, the son of a British immigrant that became an ardent Confederate and made his fortune helping the British run the Union blockade in the Gulf of Mexico during the Civil War. So, his father essentially made his fortune as a "gun runner" for the Confederacy.

After the death of his father in 1880, House ran his father's business. House also helped to make four men Governors of Texas: James S. Hogg (1892), Charles A. Culberson (1894), Joseph D. Sayers (1898) and S.W.T. Lanham (1902). After the election, House acted as unofficial advisor to each governor. Hogg gave House the title "Colonel" by appointing House to his staff. Thereafter, he was frequently known as "Colonel House".

He eventually sold his father's plantations and invested in banking. He moved to New York after 1902. There, he became associated with agents of the Rothschild bankers and J.P. Morgan crowd. He later traveled to London and was soon initiated into the Roundtable. Later, he would return to the New York and get involved with politics – particularly the campaign of Woodrow Wilson.

In 1911, prior to Wilson's taking office as President, House had returned to his home in Texas and completed a book called *Philip Dru – Administrator*. Ostensibly a fiction novel, it was actually a detailed plan for the future government of the United States, "which would establish Socialism as dreamed by Karl Marx", according to House.

This book predicted the enactment of the graduated income tax, excess profits tax, unemployment insurance, social security, and a flexible currency system. In short, it was the blueprint which was later followed by the Woodrow Wilson and Franklin D. Roosevelt administrations. House was well acquainted with the plans of the Roundtable and described them in this fiction.

Colonel House would become the closest advisor to president Wilson. House handpicked the entire Wilson Administration cabinet and virtually singlehandedly ran the State Department. When Congress passed the Federal Reserve Act of 1913, they did so without a proper quorum being present. Most Congressmen were home for the Christmas Holidays.

Also, the act was unconstitutional as it violated the provision in Article 1, Section 8 of the Constitution which states that Congress shall have the power "To coin money, regulate the value thereof…". This act would allow the Federal Reserve Bank to have that power. However, Mandell House persuaded President Wilson not to veto this unconstitutional Act.

By 1914 the Tripartite of Briton, France and Russia, was fighting a war with the central powers of Germany, Austria and the Ottoman Turks. This War was started by Austria attacking Serbia after the assassination of the Archduke of Franz Ferdinand of Austria by an ethnic Serb, who was a member of an Illuminati order. Serbia was allied with Russia and Austria with Germany. And they quickly joined in the fight which then brought in England and France on the side of Russia because of their alliances.

Also, in the background, England had already been covertly weakening the Ottoman Turks by splitting Arabia from Turkey to have access to Arabian oil. One British strategy was to create extremist Islamic religious sects like the Wahhabi and Salafi Sects by different agents of British intelligence. These sects would then be used to further British geopolitical goals.

The Wahhabi leader, Abdul Wahhab, after obtaining a sufficient following, would declare a Jihad against the Ottoman sultan at Mecca. In Egypt, the Salafis were a militant order that the British likewise created and later used to guard the Suez Canal. Also, the British sided with the Saudis, a tribe of dessert bandits, because they were enemies of the Ottomans, and would later place them in power in Arabia. (2)

Wilson had run for president by stating that he would not get Americans involved in a European war. But, there were agreements being made in England that would soon end the neutrality of the United States.

The German U-Boats had created de facto blockade of England because of the tonnage of British shipping being sunk. The war was in a stalemate and the Germans had offered a truce with England that was quite fair. Essentially both sides would stop fighting and everything would go back to the way it was before. England was ready to accept.

However, British Lord, Lionel Walter Rothschild saw an opportunity that could create a homeland for the Jewish people in Palestine, which was then controlled by the Ottoman Turks. This opportunity would disappear if England accepted Germany's peace offer. So, he engaged in urgent meetings with another member of the Roundtable, English Lord, Balfour.

In essence Lord Rothschild offered to use their influence in the United States to get the U.S. to enter the war on the side of England if in return Lord Balfour would issue the famous Balfour Declaration. When this declaration, which stated that England would provide a homeland for the Jewish people in Palestine, was made, England did not even control Palestine. But, they planned to if the U.S. would help them win the war.

Due to some opposition within the British Cabinet and by some orthodox Jewish Rabbis in Europe to the Zionist program, the document (Balfour Declaration) went through four drafts before its final form. The drafting began under the guidance of Chaim Weizmann (a Russian Zionist leader, who served as the President of the Zionist Organization) to the Zionist drafting team (Jewish Zionist Nathaniel Rothschild and Leopold Amery, and pro-Zionist Alfred Milner, Cecil Rhodes and Richard Haldane.)

Unlike Balfour Declaration that was a mere letter from a British official to a British citizen, both aliens to Palestine and had never lived in the country, Hussein-McMahon agreement of 1915-1916 was an official political agreement between the representatives of two countries that had precedence over the 1917 Balfour Declaration.

Hussein-McMahon agreement/correspondence was a series of ten letters, both in English and Arabic, exchanged between Hussein bin Ali, the Sharif (ruler) of Mecca, and Lieutenant Colonel Sir Henry McMahon, who was the British High Commissioner to Egypt representing Britain.

The ten letters expressed the British meditated and well-thought of foreign policy and commitment to Sharif Hussein to recognize an Independent Arabia "in the limits and boundaries proposed by the Sharif of Mecca" in exchange for launching an Arab Revolt against the Ottoman Empire during WWI. The agreed upon boundaries extended from Syria north to the Arabian Peninsula south, and from the Persian Gulf east to the Mediterranean west.

Considering McMahon's promises as a formal British agreement, Hussein launched the Arab revolt against the Ottoman Empire. Thomas Edward Lawrence, known as Lawrence of Arabia, was the British liaison to Hussein, through whom Britain supplied arms to the Arab Revolt.

The Arab Revolt was the British proxy army against the Ottoman Empire and many Arabs sacrificed their lives to free their country from the Ottoman rule. The Revolt was successful in liberation the whole Arabia.

Treacherous as always, the British government in May 1916 entered into a secret agreement with France known as Sykes-Picot Agreement, whereby the two countries would split the Arab World at the end of the war into two parts under their rules. The French would rule Syria, Lebanon and parts of Iraq while Britain would rule the rest of Iraq, Palestine and Jordan.

Hussein bin Ali did not accept the Sykes-Picot Agreement, the Balfour Declaration nor an independent Jewish state in Palestine, so he needed to be "removed". The British supported and armed the Saudi family to fight Hussein, who was eventually exiled to Cypress. Then, the British Placed the Saudis in power, renaming part of the region Saudi Arabia.

Not only did the Rothschilds have tremendous financial influence in the United States, as a result of lending money to figures like J.P. Morgan, David Rockefeller, and the head of Khun Loeb Bank, but that influence increased dramatically after the passage of the Federal Reserve Act. Also, the Rothschilds owned news services, like Reuters and Associated Press and could easily influence the news globally.

Three months before the U-Boat sinking of the British ship *Lusitania*, on May 7, 1915, Colonel House was meeting in London with the British Foreign Secretary, Lord Grey. Both men wanted better relations between England and the United States. The press in the U.S. started a program of vilifying the Germans after the Balfour Declaration was issued.

Even though the Germans had run full page advertisements in New York papers warning passengers not to travel on the *Lusitania* because it was carrying weapons of war to Britain and so was a legitimate war target, Germany was vilified over this incident which took over a thousand lives including 128 American lives.

The program of German vilification continued in the press. This vilification was obviously slanted reporting, as the British were also committing atrocities which the press conveniently overlooked.

At the same time, the Jews collectively joined in on the side of England and declared war on Germany. This "war' primarily took the form of a boycott of German goods. At this time, the Jews in Germany were treated better than in the rest of Europe. But with the Balfour declaration, they sided with England. This would later cause German anger towards the Jews because of their betrayal of Germany's hospitality to the Jews.

On April 2, 1917, President Woodrow Wilson, acting on the advice of Colonel House, asked a special joint session of the United States Congress for a *declaration of war* against the

German Empire. Congress responded with the *declaration* on April 6.

House played a major role in shaping wartime diplomacy. Wilson had House assemble "The Inquiry", a team of academic experts to devise efficient postwar solutions to all the world's problems. In September 1918, Wilson gave House the responsibility for preparing a constitution for a League of Nations. House helped Wilson outline his Fourteen Points and worked with the president on the drafting of the Treaty of Versailles and the Covenant of the League of Nations

In October 1918, when Germany petitioned for peace based on the Fourteen Points, Wilson charged House with working out details of an armistice with the Allies. After the war, on May 30, 1919, House participated in a meeting in Paris which laid the groundwork for establishment of the Council on Foreign Relations (*CFR*). The CFR was operational in the United States by 1921 and Mandell House is considered a founder of this organization. House also attended meetings on the Treaty of Versailles in Paris.

The Fourteen Points were promises to Germany to persuade them to surrender. However, after Germany did surrender, none of the Fourteen Points were implemented. The Treaty of Versailles was so unfair to Germany, it virtually guaranteed another war would break out. Within two decades, another war did break out – World War II!

Another feature of World War I was the use of authoritarian nationalism, characterized by dictatorial power, suppression of political opposition and control of industry and commerce, known as fascism. Fascists believe that democracy is obsolete and that a one-party as necessary to prepare a nation for armed conflict and to respond effectively to economic difficulties. Such a state is led by a strong leader—such as a dictator and a martial government to forge national unity and maintain a stable and orderly society. Fascism rejects assertions that violence is automatically negative in nature and views political violence, war, and imperialism as means that can achieve national rejuvenation.

The first fascist movements started in Italy, under Mussolini during the first world war. But later fascism spread to Spain, under Franco and Germany under Hitler. The Jesuit controlled Vatican assisted all these fascists rise to power.

In the late nineteenth and early twentieth centuries, The United States led the world in race-based lawmaking, as a broad political consensus favored safeguarding the historically white character of the country. Congress passed immigration legislation designed to guarantee the predominance of immigrants from northern Europe, largely shutting the door on Jews, Italians, Asians and others. As Nazi commentators approvingly put it, this was law intended to keep out "undesirables."

The eugenics movement took root in the United States in the early 1900's, led by Charles Davenport (1866-1944), a prominent biologist, and Harry Laughlin, a former teacher and principal interested in breeding. In 1910, Davenport founded the Eugenics Record Office (ERO) at Cold Spring Harbor Laboratory on Long Island "to improve the natural, physical, mental, and temperamental qualities of the human family". Laughlin was the first director. Field workers for the ERO collected many different forms of "data", including family pedigrees depicting the inheritance of physical, mental, and moral traits. They were particularly interested in the inheritance of "undesirable" traits, such as pauperism, mental disability, dwarfism, promiscuity, and criminality. The ERO remained active for three decades.

The Rockefeller family, among others, financed Eugenics research at the Kaiser Wilhelm Institute in Nazi Germany, where some of the most horrifying "scientific" research was conducted – including the work of Josef Mengele

In 1935, the *National Socialist Handbook on Law and Legislation*, a basic guide for Nazis as they built their new society, would declare that the United States had achieved the "fundamental recognition" of the need for a race state.

Beyond its laws, the Nazis also admired America's conquest of the West. In 1928, Hitler praised the Americans for having "gunned down the millions of Redskins to a few hundred thousand" while founding their continental empire.

And they knew that the United States had emerged as the dominant great power in the world after World War I. To them, racism had made America great. These factors added the element of racism and eugenics to create "a purified race" to the Nazi creed.

There were many industrialists and bankers in the United States that also thought fascism was a good idea. They were opposed to labor unions and communism. Obviously, the less you should pay a worker, the more profit your company can make. Corporate CEOs like to have power over others and if it was legal, they would probably favor slave labor.

In any case, many companies favored fascism during and after World War I. These companies included, Standard Oil, The Harriman Fifteen, DuPont, Ford, ITT, IBM, Union Bank, Dow Chemical and many others. And this fact would have repercussions during the great depression of the 1930s and the following war years of the early 1940s.

By the time the Council on Foreign Relations (CFR), the Federal Reserve Act., and the progressive income tax were fully operational in the United States, the Roundtable considered their goal of the peaceful return of the American colonies to British control as accomplished.

After all, many U.S. Presidents and Administration officials were members of the CFR and the CFR virtually controlled U.S. foreign policy. Also, the Federal Reserve Bank and graduated income tax allowed the Rothschild Bankers in the City of London access to the wealth of the United States. These were the main two things that the Roundtable really wanted from the United States.

This was the beginning of the World War I "special relationship" between England and the United States - two nations who had been former enemies during the Revolutionary War, the War of 1812, and the Civil War (where England sided with the Confederacy). This special relationship would continue to the present.

On September 12, 1939, the Council on Foreign Relations began to take control of the Department of State. On that day Hamilton Fish Armstrong, Editor of Foreign Affairs, and Walter H. Mallory, Executive Director of the CFR, paid a visit to the State Department. The Council proposed forming groups of experts to proceed with research in the general areas of Security, Armament, Economic, Political, and Territorial problems. The State Department accepted the proposal. The project (1939-1945) was called "Council on Foreign Relations War and Peace Studies". Hamilton Fish Armstrong was Executive Director.

In February 1941 the CFR officially became part of the State Department. The Department of State established the Division of Special Research. It was organized just like the Council on Foreign Relations War and Peace Studies project. It was divided into Economic, Political, Territorial, and Security sections. The Research Secretaries serving with the Council groups were hired by the State Department to work in the new division. These men also were permitted to continue serving as Research Secretaries to their respective Council groups. Leo Pasvolsky was appointed Director of Research.

In 1942 the relationship between the Department of State and the Council on Foreign Relations strengthened again. The Department organized an Advisory Committee on Postwar Foreign Policies. The Chairman was Secretary Cordell Hull, the vice chairman, Under Secretary Sumner Wells, Dr. Leo Pasvolsky (director of the Division of Special Research) was appointed Executive Officer. Several experts were brought in from outside the Department. The outside experts were Council on Foreign Relations War and Peace Studies members Hamilton Fish Armstrong, Isaiah Bowman, Benjamin V. Cohen, Norman H. Davis and James T. Shotwell.

In total there were 362 meetings of the War and Peace Studies groups. The meetings were held at CFR headquarters — the Harold Pratt house, 58 East Sixty-Eighth Street, New York City. The Council's wartime work was confidential.

In 1944 members of the Council on Foreign Relations The War and Peace Studies Political Group were invited to be active members at the Dumbarton Oaks conference on world economic arrangements. In 1945 these men and members of Britain's Royal Institute of International Affairs were active at the San Francisco conference which ensured the establishment of the United Nations.

In 1947 CFR members George Kennan, Walter Lippmann, Paul Nitze, Dean Achenson and Walter Krock took part in a psycho-political operation forcing the Marshall Plan on the American public. The PSYOP included a "anonymous" letter credited to a Mr. X, which appeared in the CFR magazine FOREIGN AFFAIRS. The letter opened the door for the CFR-controlled Truman administration to take a hard line against the threat of Soviet expansion. George Kennan was the author of the letter. The Marshall Plan should have been called the Council on Foreign Relations Plan. The so-called Marshall Plan and the en-suing North Atlantic Treaty Organization defined the role of the United States in world politics for the rest of the century.

These CFR members also played a large role in persuading President Truman to pass the National Security Act of 1947. This act went a long way to destroy our constitutional form of gov-ernment by creating secret agencies like the CIA, NSA, NRO and others which are virtually beyond the control of our duly elected government representatives.

In 1950 another PSYOP resulted in NSC-68, a key cold war doc-ument. The NSC (National Security Council) didn't write it — the Department of State Policy Planning Staff did. The cast of char-acters included CFR members George Kennan, Paul Nitze and Dean Achenson. NSC-68 was given to Truman on April 7, 1950. NSC-68 was a practical extension of the Truman doctrine. It had the US assume the role of world policeman and use 20% of its gross national product ($50 billion in 1953) for arms. NSC-68 provided the justification - the world-wide communist threat!

NSC-68 realized a major CFR aim — building the largest military establishment in peacetime History. Within a year of drafting NSC-68, the security-related budget leaped to $22 billion, armed forces manpower was up to a million. The following year the NSC-68 budget rose to $44 billion. In fiscal 1953, it jumped to $50 billion. Today (2017) we are still running over $800 billion dollar defense budgets despite the collapse of the USSR in 1991. Even worse, the Pentagon regularly can't account for $ Trillions of missing money.

America would never turn back from the road of huge military spending. Yet, these events and the role played by the CFR in sponsoring and carrying out the events are missing from our History books. This purposeful omission of history is what I am referring to as secret history. And interestingly, the Nazi International effort in creating the Cold War seems to correspond with the British Illuminati and their Roundtable and CFR efforts. Efforts which, I posit are coordinated via the Jesuit's Illuminati and the Jesuit Superior General at the Vatican.

In addition to the Military Industrial Complex we have a "secret intelligence industrial complex" comprised of the big five conglomerate of intelligence contractors – Leidos Holdings, CSRA, CACI, SAIC, and Booz Allen Hamilton. The work they do is "top secret and unreported."

Senator Daniel Inouye, himself blew the whistle on the shadow government during the Iran-Contra hearings in 1987. At the time Inouye expressed that the "shadow government had its own funding mechanism, shadowy Navy, and Air Force with freedom to pursue its own goals free from all checks and balances and free from the law itself."

Also, the agency known as the Joint Special Ops Command (JSOC) is the "president's secret army" which he can use for secret assassinations, overturning governments and things the

American people don't know about. So, the United States, has strayed far from its limited constitutional charter. Secrecy truly is the enemy of democracy.

An interesting company, with a large future impact, was started around the turn of the century in Saint Luis, Missouri by Knight of Malta, John Francisco Queeny. That company was Monsanto, named after John Queeny's wife, Olga Mendez Monsanto. John Queeny was also a member of the Missouri Historical Society and a director of the Lafayette South Side Bank and Trust company. Monsanto started producing the artificial sweetener, saccharin for the Coca Cola company.

Soon, Monsanto diversified into phenol (a World War I -era antiseptic), and aspirin when Bayer's German patent expired in 1917. Monsanto began making aspirin, and soon became the largest manufacturer world-wide. Bayer, the German competition, cut prices, in an effort to drive Monsanto out of business, but failed.

During world war I, they were unable to import chemical products, primarily from I.G. Farben, the German chemical giant. So then, Monsanto started manufacturing their own. They also produced phosphorus, Vanillin, caffeine, sedative drugs, laxatives and later, the now outlawed, pesticide DDT. In the 1920's, Queeny's son, Edgar Queeny, took over and built Monsanto into a global powerhouse, extending into the production of an astounding array of plastic, rubber and vinyl goods, fertilizers, herbicides and pesticides. We will hear more about Monsanto later in this account.

Another development was the Creation of the Federal Bureau of Investigation (FBI). Its first incarnation started as the National Bureau of Criminal Identification in 1896.

This was reorganized in 1908 and called the Bureau of Investigation (BOI). Knight of Malta, John Edgar Hoover was appointed as the fifth director of the BOI in 1924. Hoover then reorganized the BOI into a new bureau in 1935, called the Federal Bureau of Investigation (FBI).

In the FBI, Hoover has been credited with building the FBI into a larger crime-fighting agency than it was at its inception, and with instituting a number of modernizations to police technology, such as a centralized fingerprint file and forensic laboratories.

Hoover became a controversial figure as evidence of his secretive abuses of power began to surface. He was found to have exceeded the jurisdiction of the FBI, and to have used the FBI to harass political dissenters and activists, to amass secret files on political leaders, and to collect evidence using illegal methods, such as illegal wiretaps.

According to President Truman, Hoover transformed the FBI into his private secret police force. Truman stated: "we want no Gestapo or secret police. The FBI is tending in that direction. They are dabbling in sex-life scandals and plain blackmail". Richard Nixon was recorded as stating in 1971 that one of the reasons he did not fire Hoover was that he was afraid of reprisals against him from Hoover.

So, the Jesuit method of collecting and using files on people for blackmail and other purposes was placed into the FBI by Knight of Malta, John Edgar Hoover.

In the 1920s in Germany, three secret societies were working together to create flying saucers. These secret societies were allied by a metaphysical understanding. These Societies were the Vrill Society, the Thule Society and the Swartz Sonne (Black Sun).

The Vrill Society was organized by Maria Orsic and was originally comprised of female psychics. The Thule Society was formed by Rudolph von Sebottendorf and was even deeper into the Metaphysical. Many prominent Germans were members. The Swartz Sonne was a secret inner circle of the Thule Society.

Maria Orsic channeled information from beings that claimed to be from the star system Aldebaran which included plans for the Jensietsflugmaschine (JFM, in English, Other World Flying Machine). The JFM resembled the flying saucers that many have seen. Neither Maria Orsic or her father understood the technical side of these plans.

However, Maria's father knew Professor Winfried Otto Schumann at the Technical University at Munich and took the channeled plans to him. Dr. Schumann became quite interested in the plans and decided to build one.

Maria Orsic was also in contact with the Thule Society, containing a number of influential and financially independent persons. The Thule Society decided to provide the financing to build the first saucer, JFM1, after consulting with Dr. Schumann.

On March 22, 1922, the first version of this saucer was tested. It lifted about 50 feet off the ground and started to violently wobble, the pilot brought it down and jumped out of the ship before it broke apart. Maria Orsic did some more channeling and the engineers went back to the drawing boards.

A second version called the JFM2 was tested on December 17, 1923. It was a remote-controlled version and successfully flew for 55 minutes. This was the beginning of the Vrill levitation disks, which were developed much further. This remarkable technology was kept secret from the German people and the world at large. In German secret societies, Dr. Schumann was credited as being the father of the Vrill levitation disks. In the U.S. he is merely noted for discovering the Earth's Schumann resonance.

Later with the event of World War II, the Nazi SS discovered this project and took the technology from the Vrill Society and embarked on their own secret flying saucer development for weapons of war.

This is only a brief covering of this subject. Much more detail is covered in my book referenced in the bibliography foot note. We will examine more on the future development of this subject later in this account. (3)

Some financiers and industrialists actually were pro-communist. These were persons that wanted monopolies. What better monopoly than one run by the state and you controlled the state?

William Lawrence Saunders, chairman, Ingersoll-Rand Corp.; director, American International Corp.; and deputy chairman, Federal Reserve Bank of New York on October 17, 1918 wrote a telling letter to U.S. President, Woodrow Wilson. The following is an excerpt from the beginning of this letter:

"Dear Mr. President:

I am in sympathy with the Soviet form of government as that best suited for the Russian people ..."

This was used as an introduction to Antony C. Sutton's book, *Wall Street and the Bolshevik Revolution: The Remarkable True Story of the American Capitalists Who Financed the Russian Communists.*

While monopoly control of industries was once the objective of men like J. P. Morgan and J. D. Rockefeller, by the late nineteenth century the inner sanctums of Wall Street understood that the most efficient way to gain an unchallenged monopoly was to do it by political means.

One theory of communism held that the state should control the means of production which in theory would provide the needs of the people. The ideal monopoly then, would be to control the communist state that would control everything else. The proper indoctrination could make society go to work for the monopolists— under the name of the public good and the public interest. This strategy was detailed in 1906 by Frederick C. Howe in his *Confessions of a Monopolist.*

The Marburg Plan, financed by Andrew Carnegie's ample heritage, was produced in the early years of the twentieth century. The governments of the world, according to the Marburg Plan, were to be socialized while the ultimate power would remain in the hands of the international financiers. This idea was knit with other elements with similar objectives. Lord Milner of the British Roundtable, in England provides the transatlantic example of banking interests recognizing the virtues and possibilities of Marxism. Ivy Lee, Rockefeller's public relations man, made similar pronouncements, and was responsible for selling the Soviet regime to the gullible American public in the late 1920s.

In New York the socialist "X" club was founded in 1903. It counted among its members not only the Communist Lincoln Steffens, the socialist William English Walling, and the Communist banker Morris Hillquit, but also John Dewey, James T. Shotwell, Charles Edward Russell, and Rufus Weeks (vice president of New York Life Insurance Company). The annual meeting of the Economic Club in the Astor Hotel, New York, also witnessed socialist speakers.

From these unlikely seeds grew the modern internationalist movement, which included not only the financiers Carnegie, Paul Warburg, Otto Kahn, Bernard Baruch, and Herbert Hoover, but also the Carnegie Foundation and its progeny International Conciliation. The trustees of Carnegie were also prominent on the board of American International Corporation (AIC).

In 1910, Carnegie donated $10 million to found the Carnegie Endowment for International Peace, and among those on the board of trustees were Elihu Root who headed the Root Mission to Russia, in 1917, Cleveland H. Dodge (a financial backer of President Wilson), George W. Perkins (Morgan partner), G. J. Balch (AIC and Amsinck), R. F. Herrick (AIC), H. W. Pritchett (AIC), and other Wall Street luminaries.

Woodrow Wilson came under the powerful influence of—and indeed was financially indebted to—this group of internationalists. And many of these Internationalists were also members of Illuminati organizations like Skull and Bones and the British Roundtable. As Jennings C. Wise has written, "Historians must never forget that Woodrow Wilson . . . made it possible for Leon Trotsky to enter Russia with an American passport."

The October Revolution of 1917—in which Bolshevik communists led by Vladimir Lenin seized power in Russia—also greatly influenced the development of fascism - to counter communism. This was a typical strategy of divide and conquer often used by the Jesuit's Illuminati.

This Bolshevik Revolution was largely financed by two banks that were Class A shareholders in the newly created Federal Reserve Bank consortium – The Khun Loeb Bank of New York and the Max Warburg Bank of Hamburg, Germany. This revolution completed the Illuminati purposes of World War I, as outlined by Albert Pike 3 decades earlier. (4)(5)

The Great Depression

The 1920s was a time of low interest and easy credit and, in general the economy in the United States was rapidly growing. Low interest loans could be used to start businesses, buy farms and homes which kept increasing in value. The stock market was also increasingly seen by the public, as a way to become rich, with stocks continually raising in value.

Also, Prohibition was a period of nearly 14 years of U.S. history (1920 to 1933) in which the manufacture, sale, and transportation of intoxicating liquor was made illegal by an amendment to the constitution.

It was a time characterized by speakeasies, glamour, and gangsters and a period of time, in which even the average citizen broke the law. Organized crime also got a boost from Prohibition as the smuggling of outlawed liquor became quite profitable. Al Capone made huge profits from "Rum Running." Others also joined in this profitable enterprise, like Joseph Kennedy, the father of President John F. Kennedy, who made his fortune Rum Running. So, Prohibition also boosted the economy in an unforeseen way.

Interestingly, Prohibition, sometimes referred to as the "Noble Experiment," led to the first and only time an Amendment to the U.S. Constitution was repealed.

This booming economy suddenly ended in 1929, as was planned by the Illuminati bankers. They had invented a financial product called the "24-hour call loan." Somewhat like Goldman Sachs Bank (another Class A stockholder in the Fed) created credit default swaps, mortgage backed securities and other exotic derivatives.

As the Stock market kept rising, many would purchase stock on borrowed money. The brokers offered these 24-hour call loans to purchasers of stock on margin. Some of these margins would be as high as 90%, where the borrower only put up 10% and the broker 90% to buy the stock. The contractual understanding was that if the broker called the loan, it would have to be repaid within 24 hours. This usually meant the borrower would have to quickly sell the stock to repay the loan. Those hoping to get rich quick weren't worried because the stock prices only seemed to keep increasing and they would probably profit from the sale.

The Dow Jones Industrial Average soared from 85.76 on October 27, 1923 to, at the time, an all-time high of 381.17 on September 3, 1929. Fifty-five days later, on Monday, October 28, and continuing on Black Tuesday October 29, 1929, the stock market collapsed. This was caused by the brokers all calling in their loans at the same time, forcing a massive sell off of stocks. Even more conservative investors who owned stock that wasn't on margin were inclined to sell, to cut their losses short.

In coordination with the Wall Street brokers, the Federal Reserve Bank raised the interest rates while reducing the money supply. Similar strategies were used in Europe and a global depression soon followed. Speculators lost their shirts; banks failed; the nation's money supply diminished; and companies went bankrupt and began to fire their workers in droves.

The unemployed were unable to pay the mortgage payments on their homes, which were repossessed. Likewise, many mortgaged farms were also repossessed.

Meanwhile, President Herbert Hoover urged patience and self-reliance: He thought the crisis was just "a passing incident in our national lives" that it wasn't the federal government's job to try and resolve. By 1932, one of the bleakest years of the Great Depression, at least one-quarter of the American workforce was unemployed.

The Federal Reserve Bank was so successful at draining the wealth of the United States, that it bankrupted it by 1933, only two decades after its creation. One of the ways that this was done was by loaning Federal Reserve Notes into circulation at interest – primarily to the government. There were not enough notes around to pay back the debt plus interest. Hence, the national debt would keep increasing.

These notes were redeemable in gold but more notes were issued than the Fed. actually possessed in gold. This was based on fractional banking where, over the years, banks realized that depositors rarely removed more than 10% of their savings and the banks could safely loan out more than they possessed. Usually banks could loan out 10 times their actual holdings in gold in the form of paper notes to increase their profits. So, if they loaned out ten times more than their investment in gold holdings at 5% a year. They would really make 50% per year on their actual investment using fractional banking!

President Franklin Delano Roosevelt placed the United States in a state of emergency and removed the gold backing of the U.S. Dollar.

During a "Fireside Chat" on 07 May 1933, Roosevelt basically admitted that gold-clause obligations far exceeded the amount of gold held by the US Treasury and Federal Reserve. In fact, the total gold obligations far exceeded the amount of gold in the entire world! This is why FDR had to remove the gold backing of the U.S. Dollar.

All persons possessing gold were required by law to turn it in to banks at $20.67 per ounce. This caused the gold reserves of the US Treasury and Federal Reserve to increase. *After* most of the public's gold was turned in, FDR raised the official price from $20.67 to $35.00 per troy ounce. This increased the Fed banker's holdings even further, at the expense of ordinary citizens who had previously owned gold.

When President Franklin Roosevelt took office in 1933, he acted swiftly to try and stabilize the economy and provide jobs and relief to those who were suffering. Over the next eight years, the government instituted a series of experimental projects and programs, known collectively as the New Deal, that aimed to restore some measure of dignity and prosperity to many Americans.

More than that, Roosevelt's New Deal permanently changed the federal government's relationship to the U.S. populace, as predicted in Mandel House's book, *Philip Dru – Administrator*.

About the same time that FDR came into power in the U.S., Hitler rose into power in Germany. Germany's depression was much worse than in the United States.

Unemployment in the Weimar Republic of Germany was around 50%! On top of this there was extreme inflation, which got so bad that it literally took a wheelbarrow full of German marks to shop for groceries!

Assisting Hitler's rise to power were the Vatican, international bankers, and American industrialists, who believed that fascism was not only the solution to Germany's problem but, perhaps that of other countries as well.

On the American contribution to Hitler's rise to power, one excellent historian, Anthony Sutton had this to say:

> "The contribution made by American capitalism to German war preparations can only be described as phenomenal. It was certainly crucial to German military capabilities.... Not only was an influential sector of American business aware of the nature of Nazism, but for its own purposes aided Nazism wherever possible (and profitable) - with full knowledge that the probable outcome would be war involving Europe and the United States."

Professor Sutton reveals one of the most remarkable and under-reported facts of World War II -that key Wall Street banks and American businesses supported Hitler's rise to power by financing and trading with Nazi Germany. Carefully tracing this closely guarded secret through original documents and eyewitness accounts, Sutton comes to the unsavory conclusion that the catastrophe of World War II was extremely profitable for a select group of financial insiders. He presents a thoroughly documented account of the role played by J.P. Morgan, T.W. Lamont, the Rockefeller interests, General Electric, Standard Oil, and the National City, Chase, and Manhattan banks, Kuhn,

Loeb and Company, General Motors, Ford Motor Company, and scores of others in helping to prepare the bloodiest, most destructive war in history. (6)

With all this help, Hitler soon had full employment and a sound currency in Germany. Not only did he order the creation of the famous Autobahn, but put Germans to work creating the German war machine formerly prohibited by the Treaty of Versailles. In Italy, Mussolini was creating a similar economic miracle. Both countries quickly left their depressions behind.

This Hitler economic miracle in Germany and the Mussolini economic miracle in Italy impressed many businessmen on both sides of the Atlantic. A number of these men in the U.S. believed that fascism – not communism or socialism – would be the solution to the depression in the United States. These men were virulently opposed to FDR's New Deal policies, which they felt were too communistic, and started to plan a military coup against President Roosevelt.

In November 1934, Marine Gen. Smedley Butler gave secret testimony before the McCormack-Dickstein committee – a precursor to the House Committee on Un-American Activities. In it, Butler told of a plot headed by a group of wealthy businessmen to establish a fascist dictatorship in the United States, complete with concentration camps for "Jews and other undesirables."

World War I hero, Marine General Smedley Butler had been approached by Gerald P. MacGuire of Wall Street's Grayson M-P Murphy & Co. MacGuire claimed they would assemble an army of 500,000 mostly unemployed WWI veterans and march on DC. McGuire's group wanted Butler to lead the coup.

They promised to put up $3 million as starters and dangled a future $300 million as bait. Butler pretended to go along with the plot until he could learn the identities of all the schemers.

McGuire's group was part of *The American Liberty League*. This League was headed by the DuPont and J.P Morgan cartels and had major support from Andrew Mellon Associates, Pew (Sun Oil), Rockefeller Associates, E.F. Hutton Associates, U.S. Steel, General Motors, Chase, Standard Oil and Goodyear Tires.

Money was funneled thru the Senator, Prescott Bush-led Union Banking Corporation and the Prescott Bush-led Brown Brothers Harriman to *The American Liberty League*. The plotters bragged about Bush's Hitler connections and even claimed that Germany had promised Bush that it would provide materiel for the coup.

This claim was entirely believable because a year earlier, Chevrolet president William S. Knudsen (who himself had donated $10,000 to *The American Liberty League*) went to Germany and met with Nazi leaders and declared upon his return that Hitler's Germany was "the miracle of the twentieth century." At the time, GM's wholly-owned Adam-Opal Co. had already begun producing the Nazi's tanks, trucks and bomber engines. James D. Mooney, GM's vice-president for foreign operations was joined by Henry Ford and IBM chief Tom Watson in receiving the Grand Cross of the German Eagle from Hitler for their considerable efforts on behalf of the Third Reich.

The military coup plotters were not very good at sizing up Smedley Butler, who had become totally disgusted with what he called the war racket. If they had been, they would have approached a different person to head their plot.

On August 21, 1931, Butler had given a speech to an American

Legion convention in New Britain, Connecticut. Here are excerpts from that speech:

> "I spent 33 years...being a high-class muscle man for Big Business, for Wall Street and the bankers. In short, I was a racketeer for capitalism....
>
> I helped purify Nicaragua for the international banking house of Brown Brothers in 1909-1912. I helped make Mexico and especially Tampico safe for American oil interests in 1916. I brought light to the Dominican Republic for American sugar interests in 1916. I helped make Haiti and Cuba a decent place for the National City [Bank] boys to collect revenue in. I helped in the rape of half a dozen Central American republics for the benefit of Wall Street....
>
> In China in 1927 I helped see to it that Standard Oil went its way unmolested.... I had...a swell racket. I was rewarded with honors, medals, promotions.... I might have given Al Capone a few hints. The best he could do was to operate a racket in three cities. The Marines operated on three continents..."

While the Committee found that Gen. Butler was telling the truth, a program of discrediting Butler was started by the elite financiers behind the plot. Quickly, the corporate press weighed in and sought to raise doubts about the war hero, settling on branding him naive. Knudsen's meme was: "it was all idle cocktail party chatter." This red herring was trumpeted under the Associated Press headline "The Cocktail Putsch." New York Mayor Fiorella LaGuardia dismissed the plot as "someone at the party had suggested the idea to the ex-Marine as a joke." From 1934 through 1936, *The American Liberty League* got thirty-five pro-League front page stories in the *New York Times*.

TIME ridiculed Butler in a Dec. 3, 1934 cover story, even though Butler's story was corroborated by Veterans of Foreign Wars head, James E. van Zandt, who said he was also approached to lead the coup.

So, there was a definite pro-Nazi support among certain U.S. industrialists and bankers during this time in history – and some of this support amounted to treason! What happened to these treasonous coup plotters, you might wonder? Absolutely nothing! Like banks that are too big to fail, these criminals were too big to jail. Congress did not even bother to subpoena any of them for questioning. The news media and history books swept the whole affair under the carpet.

World War II

The Vatican assisted the rise to power of all three of the fascists of Europe. Mussolini was raised to power in Italy via the tremendous influence of the Catholic Church in Italy. In return, Mussolini signed the Lateran Treaty of 1929, which gave the Vatican sovereign status, outside the jurisdiction of Italy and returned $85 million to the Vatican for the many formerly confiscated properties of the Catholic Church. This treaty also made Catholicism the only recognized religion in Italy.

Franco was also supported by the anti-communist, Catholic Church during the Spanish Civil War. Without the support of the Vatican, Franco would not have won that war and be placed into power in Spain.

Likewise, without the support of the Vatican, Hitler would never have come into power. The Vatican Secretary of State, Eugineo Pacelli ordered the dissolution of the Catholic Centrum party in Germany – the only political party which could have defeated Hitler in the elections that placed him into power in 1933.

A Concordant between the Vatican and the Nazi party was worked on for years by Pacelli who also oversaw the signing of this Concordant. Pacelli would later become Pope Pious XII. This Concordant basically kept the Catholic Church out of German politics while strengthening the Vatican's control over German Catholics. So, even though many German Catholics were opposed to Hitler, the Vatican removed their power to democratically oppose him. Also, this was another manifestation of the Jesuit control over the Vatican.

The Jesuit, Himmler bragged that in creating the Nazi SS, he had patterned it closely after the Jesuit's Society of Jesus.

So, Adolf Hitler's rise to power was also greatly assisted by the Vatican as thoroughly documented in Avro Manhattan's *The Vatican's Holocaust* and Edmond Paris' *The Secret History of the Jesuits.*

The Book *Mien Kampf* (my Struggle) supposedly written by Hitler was actually ghost written by the Jesuit priest, Father Staempfle. In many German Churches, *Mien Kampf* was placed alongside the bible on the church alters. Hitler was himself chosen to lead Germany by the Jesuit Superior General, Vladimir Ledochoski, because of Hitler's oratory and leadership ability.

In 1922, "Catholic Action" was created by Pope Pious XI on the advice of his two Jesuit Confessors, Father Alissiardi and Father Celebrano. Catholic Action had the mission of getting Catholics involved in political activity. Often radio and film was used to promote Catholic Action agendas. This agenda was strongly anticommunist and promoted fascism as the guard against communism in Europe.

In France, and Czechoslovakia, Catholic Action acted as a "fifth column" to assist the German invasion of those nations. After the invasion, Catholic Action films in Vichy France persuaded many Frenchmen to volunteer to join the German army on the Eastern Front against the "bloody Bolsheviks". After the war, Catholic Action organized many charitable organizations to help the poor, in order to counter the communist influence.

After Hitler's invasion of West Poland in 1939, England declared war on Germany because of treaty agreements between Poland and England. Interestingly, Soviet Russia later invaded East Poland, but in that case, the alliance between the Soviets and England was unaffected by this same treaty.

Hitler really did not want another war with England – a country he had a lot of respect for. When English troops in France were evacuated at Dunkirk, they were sitting ducks for Germany's Luftwaffe and artillery. But, Germany allowed the evacuation without interfering.

It was only after Winston Churchill became Prime Minister of England after Chamberlain resigned, that the war between England and Germany would quickly escalate. Many considered Churchill even more of a war monger than Hitler.

I don't intend to cover the history of World War II here. However, I do wish to touch on some highly remarkable secret operations that occurred during this war, and the coverup of U.S. corporations actively helping Germany during the war.

Of all the intelligence agencies in the U.S. during World War II, Naval Intelligence was considered the best. But, there was even a more secret intelligence operation that was hidden from Naval Intelligence. This operation was headed by Admiral Rico Botta working at North Island, San Diego, California.

His daytime job was officially stated to be a Naval aircraft maintenance officer at the Naval Air Station at North Beach. At night, Botta was covertly managing an operation deep inside Nazi Germany with 29 specially trained operatives. These operators were U.S. citizens of German background, who could speak fluent German and had penetrated the Nazi SS secret flying saucer projects in Germany.

These operators could hardly believe the information they were discovering! The Nazi SS, under Hans Kammler, not only was using antigravity technology developed by scientists like, Dr. Schumann and Victor Schauberger but also used the work of other scientists in other countries, like Thomas Townsend Brown and Nikola Tesla to develop 30 different versions of flying saucers, they also had made treaties with extraterrestrials and intraterrestrials, known as the Draco Reptilians, which had also had given them their own fully manufactured and functioning saucers and cigar shaped motherships acting as carriers for these saucers in outer space.

Hitler, himself had meetings with the leader of these Reptilians in an underground cave. Hitler referred to this Draco leader as a "superman" and said that "his eyes were terrible" and "I was afraid". This superman was directing Hitler to conquer the world in alliance with them. In return Hitler would receive their advanced weaponry and the right to create a base on the Moon.

They also showed the Germans an abandoned Reptilian underground city in Antarctica that was accessible by U-boat that the Germans could use to create a secret base. This base, built by refurbishing the abandoned city, largely with slave labor, would be called New Berlin and was under the area known to the Germans as Nue Swabenland also known as Queen Maud land.

Occasionally, one or more of these operators would return to

North Island from Germany with photographs and plans for these flying saucers and other "wonder weapons" and other stories and documentation and secretly meet at night with Rico Botta. Also attending these meetings was William Tompkins, a young naval officer with a photographic memory. Tompkins job was to design intelligence briefing packets from the debriefing of these operators. Tompkins also had to deliver these packets to various aviation companies, universities and think tanks for evaluation.

From 1942 to 1945 Tompkins had attended over a thousand debriefings of these operators that had been embedded in different German companies involved in the flying saucer development.

Apparently, there were two different developments, one in Nazi occupied Europe that were primarily developing flying saucers for weapons of war, the other under the Antarctic ice in New Berlin, that was developing these saucers that could operate underwater, in the air and be used for space travel. (7)

By 1942, the Germans had traveled from their Antarctic base to the moon and started developing a base there, with the Draco Reptilians's permission.

The Third Reich was also working on an atomic bomb. It is known that Hitler pursued the goal of nuclear technology and wanted his V-2 rockets to be able to carry them to destroy the UK. Recently declassified file APO 696 from the National Archives in Washington is a detailed survey of how far Third Reich scientists got in the development of an atomic bomb.

In the file, obtained by the popular daily newspaper Bild, the task of the academics who prepared the paper between 1944 and 1947 was the "investigations, research, developments and practical use of the German atomic bomb." The report was pre-pared by countless American and British intelligence officers and also includes the testimony of four German experts - two chemical physicists, a chemist and a missile expert. It concurs that Hitler's scientists failed in the quest to achieve a break-through in nuclear technology.

But, that conclusion may have been erroneous. A documented test may have taken place of a rudimentary atomic warhead in 1944. The statement of the German test pilot Hans Zinsser in the file is considered evidence. The missile expert says he ob-served in 1944 a mushroom cloud in the sky during a test flight near Ludwigslust.

His log submitted to the Allied investigators reads: "In early October 1944 I flew away 12-15km from a nuclear test station near Ludwigslust (South of Lübeck).

"A cloud shaped like a mushroom with turbulent, billowing sections (at about 7,000m) stood, without any seeming con-nections over the spot where the explosion took place. Strong electrical disturbances and the impossibility to continue radio communication as by lighting turned up."

The pilot described seeing a mushroom cloud during the test. He estimated that the cloud stretched for 9km and described further "strange colorings" followed by a blast wave which translated into a "strong pull on the stick" - meaning his cockpit control stick connected to the wing ailerons and tail rudder and elevator controls.

An hour later a pilot in a different machine took off from Ludwigslust and observed the same phenomenon. According to other archival documents, the Italian correspondent Luigi

Romersa observed, on the ground, the same phenomenon. He had been sent by dictator Benito Mussolini to watch the test of a "new weapon" of the Germans. He was ordered to report his impressions back to Mussolini.

The testimony of the four German scientists in the declassified American report mentions a top-secret meeting held in Berlin in 1943 at which armaments minister and Hitler favorite, Albert Speer was present for the discussion called a "nuclear summit."

In the end, the report states that the Allies believe the Germans fell short of triggering the nuclear chain reaction necessary to trigger a nuclear blast - but none could come up with an explanation for what occurred in the skies over Ludwigslust in 1944.

Did Hitler succeed in obtaining a nuclear weapon? This could have been a prototype test similar to the U.S. Trinity test. But fortunately, more A-bombs might not have been able to be produced by the Germans before the war ended in May of 1945.

In my opinion, the Germans indeed had the know-how and technology to produce an A-weapon by this time. This opinion is strongly reinforced by the unusual cargo aboard the surrendered German U-boat, U-234 in May of 1945. Part of the cargo, originally destined for Japan, was U235 or refined, bomb grade, uranium stored in gold tubes.

Some have suggested that this cargo was harmless U238. However, U235 is highly corrosive and easily contaminated and degraded by contact with other materials, hence the noble metal of gold used to store the purified U235, which was immensely more valuable than gold. Normal uranium or U238, which is not bomb grade, could easily and safely be stored steel drums.

U-234's cargo was unloaded by a Major Vance of the Army Corps of Engineers. And the Manhattan Project, charged with developing the U.S. atomic bomb, was under the Army Corps of Engineers! A coincidence? I think not. This unusual cargo was a desperate attempt to give Germany's ally, Japan, a chance to produce A-bombs before they too were defeated.

As a matter of fact, Japan was also pursuing a nuclear weapon program during World War II. The German spy, Alcazar de Velasco had already informed Germany and Japan of the Manhattan Project to create an atom bomb in 1943.

He had also hired two Mexican youths to obtain soil samples from Los Alamos in an area where a uranium trigger explosion was tested. The trigger explosion was designed to implode Uranium 235 into a critical mass causing a nuclear detonation – but itself was not an atomic explosion. The spies however, had no way of knowing this about an explosion that was itself, huge.

These soil samples were placed in glass jars and the Mexican youths and soil samples were taken to Japanese agents at the Majestic Hotel in Mexico City. The Japanese agents were, at first, more interested in the youths than the samples.

The youths were given a complete medical examination. Nails, skin and blood samples were taken and examined. The agent would not go near the soil samples until they were deposited in special shielded bags.

That was when Alcazar de Velasco realized they were checking for radioactivity exposure and wondered if he had been exposed, as he had been carrying the bottled soil samples. White blood cell count was a means of determining if a person suffered from excessive radioactive exposure.

However, the white cell count was normal in these youths and the Japanese agent concluded that the U.S. was not that far along in their progress towards a working A-bomb.

In any case, the Manhattan Project was not as secret as the U.S. had hoped it was. Both Germany and Japan were well informed of it. This realization pushed the Japanese atomic bomb effort further along.

The Japanese called their A-bomb Genzai Bakudan. They started their project in Japan. But, when the B-29 air raids started devastating their cities, they moved their atom project to the Konan area of North Korea near a hydroelectric dam where plenty of electric power was available. The lab was set up in Hungnam and Rikizo Takashai and Tadashiro Wakabayashi were the lead scientists of this project.

Their A-bomb was completed in 1945. In a deserted island in the Sea of Japan the bomb was tested just before dawn on August 12. It was a blast as powerful as the ones at Nagasaki and Hiroshima. But, it was too late for Japan because it was 6 days after the Hiroshima blast and 3 days before Japan's surrender.

Soon after this event, the Russians overran the Hungnam facility and captured the lead scientists who were hurriedly trying to destroy scientific papers explaining the nuclear process. They also captured much of the equipment used by the Japanese. And, not too many years later, the Russians also had an atomic bomb.

Much more detail of this interesting information is revealed in Robert Wilcox's, well documented and researched book, *Japan's Secret War.*

The Russians also obtained refined Uranium 235 and plans on making an atom bomb from U.S. Operation Lend Lease run by Harry Hopkins. This is according to a Lend Lease operation manager in Montana, Major Racey Jorden, as revealed in his book *From Major Jordan's Diaries*.

The U.S. was also secretly working on exotic technology of its own. The top brass of the Navy was working on Project Rainbow. This project was called Rainbow because of its magical seeming possibilities which covered a spectrum of possibilities which included controlling the weather using a Orgone energy experimental science developed by Wilhelm Reich, and making objects disappear via trans-dimensional travel using experimental science developed by Nikola Tesla.

Project Rainbow started in the mid-1930s and lasted until the Mid-1940s. The top Brass of the Navy knew that there was a planned confrontation with Germany and Japan and that the German science and technology was decades ahead of U.S. technology so they embarked on this project to counter the German technological advantage.

Alfred Bielek, who claims to have worked on the project was an early source of this secret information. His role in project was as a crewman, operating the electronic controls, under the different identity of Edward Cameron, on the Destroyer Escort, *Eldrige*, which was rigged up to disappear. The more popular known name for this classified experiment was the Philadelphia Experiment because it took place in the Philadelphia Naval Yard.

According to Bielek, Project Rainbow had a number of high level scientists and mathematicians working on the project. The list included Albert Einstein, Nikola Tesla, Thomas Townsend

Brown and Joseph von Nuemann.

The theoretical work of the mathematician David Hilbert was also used. Hilbert extended the mathematical concepts in three-dimensional space to spaces with higher dimensions generally known as a Hilbert space. In 1926, von Neumann showed that if atomic states were understood as vectors in Hilbert space, then they would correspond with both Schrödinger's wave function theory and Heisenberg's matrices.

Edward Cameron attended Princeton University and met Joseph von Neumann there. After receiving a PHD in Physics, Cameron was recruited by von Nuemann into Project Rainbow in 1939. There, Nuemann instructed Cameron how gravity, quantum physics and time really work, which was classified science not covered in the Universities.

The *Eldrige* did disappear in the experiment because it was teleported to another location in time and space. Tesla had previously warned that the powerful magnetic fields involved could have harmful effects on the crew members and wanted more time to make the experiment safer. The Naval officer in charge stated that it was a war time emergency and the experiment would go ahead anyway.

This was because German U-boats were taking a heavy toll on ships at sea and the Navy was hoping to provide this technology to make their ships disappear and not be seen by these U-boats. When the Navy disregarded Tesla's advice, Tesla resigned from the project.

When the *Eldrige* finally reappeared, total chaos reigned on board. Some of the crew were embedded in the steel of the ship. Some were on fire. Others had gone completely crazy.

Some had completely disappeared. The technical part of the experiment worked. The human part of the experiment did not. At this point, the Navy cancelled the project.

There were, however, persons that wanted to continue with more experimentation and research on this experiment gone wrong because of the exciting new scientific and technical possibilities it promised. And it continued as Project Phoenix with Joseph von Nuemann as Director.

Eventually after the bugs were worked out, safe and workable teleportation and time travel was developed by Project Phoenix. There however were still phycological effects on people involved because adult humans have been programed to consider it all impossible. It was discovered that pre-adolescent children could handle time-travel and teleportation physiologically much better than adults.

Andrew Basiago's father was involved in an offshoot of Project Phoenix called Project Pegasus. In this project, young children were trained to become chronauts that traveled in time and space and later, as properly conditioned adults, they would do the same.

According to consultant engineer for DARPA's Project Pegasus, Robert W. Beckworth, the Ralph M. Parsons Company was involved with Project Pegasus.

In 1967, Young Andrew Basiago became one of these young chronauts. He was sent back in time to the Gettysburg Address by President Lincoln and later to the Ford Theater during Lincoln's assassination. He also was teleported along with other youngsters from New Jersey to Santa Fe, New Mexico. During these teleportation's no receiver was necessary.

Later in 1980, Andrew Basiago was trained to be teleported to a secret base on Mars along with nine other persons which included Berry Soetoro, aka Barak Obama, William Stillings, Cortney Hunt, Regina Dugan and Bernard Mendez. The training course was held at the College of the Siskiyous, near Mount Shasta and taught by Major Ed Dames. Barak Obama and Ed Dames both deny being in the project.

Starting in 1981 these trainees would be sent to Mars via a "Mars Jump Room" located at 999 N. Sepulveda Blvd. in El Segundo California in the old Howard Hughes Building next to Los Angeles Airport. More details of these projects can be found in my book, *Secret Science and the Secret Space Program*.

So, the Americans had their own secret, advanced technology projects that started before World War II.

By 1943, after the defeat of the best of the German army at Stalingrad by the Soviets, and setbacks for General Rommel in North Africa at the hands of the Americans and British, the leaders of Germany knew that they would lose the war – while their propaganda machine still tried to convince the German people otherwise.

Plan "B" was decided upon by the German leaders. The strategy would thereafter be to fight a holding action to delay their eventual defeat as long, as possible. This strategy would give them time to move their best men, technology, gold and industrial secrets to neutral countries like Switzerland, Spain, Argentina and their secret base in Antarctica.

Then, after the war they would start over again, with their superior industrial secrets and looted gold. Their future takeover would be not so much by destructive war – but by outsmarting their opponents in the field of business, technology and espionage.

Many German factories were moved to huge underground facilities, like Nordhousen, where the German V-2 rockets were being assembled by slave labor. This was because the Allies were bombing the hell out of some, but not all, of the surface factories. Large freight carrying, U-boats were constructed and their best technology was transported by giant U-boats to their secret base under the ice of Antarctica.

At the war's end, most of their flying saucers were either flown away to secret bases or destroyed to prevent the Allies from recovering their secrets. Also, many who worked on this technology were killed to prevent them from talking.

By the war's end, most of Europe had been reduced to rubble. This wholesale destruction came not at the hands of the Nazi but at the hands of the allied air command. Yes, the Nazi war machine had been destroyed on the battlefield but at what cost? In Germany, it was estimated that as much as 80% of all housing units had been destroyed. The major city of Dresden was simply erased in a massive firestorm by the bombing campaign. Estimates of civilian deaths in Dresden are generally accepted at approximately 70,000. However, some suggest that the estimate is as high as 500,000. The deliberate targeting of civilians is considered a war crime.

However, the effect of the air campaign against industrial centers and munitions makers was quite different. The Nazi war machine was producing more planes, tanks, trucks, etc. at the end of the war than in 1941. Overall, production of munitions at the end of the war was estimated to be at roughly 80% of capacity.

In short, the allied bombing campaign, while targeting civilians was a highly successful war crime, the attempt to damage German war production had been pretty much a failure. The people making up the target lists for the allied bombers should have been investigated!

At the cabinet level, the air force was under the control of Secretary of War Stimpson. This Skull and Bones member advocated an "easy peace" with Germany at the end of the war. Roosevelt allowed Stimpson to choose his own staff. He chose John McCloy to act as assistant Secretary of War in charge of intelligence, civilian affairs and general troubleshooter. Stimpson placed Robert Lovett as assistant secretary of war for air. Both McCloy and Lovett had Wall Street backgrounds. McCloy had been a former Wall Street lawyer and Lovett a partner and close friend of Prescott Bush at Brown Brothers and Harriman.

It was Prescott that selected Lovett for membership in the Skull and Bones. Lovett was a fervent advocate of terror bombing of population centers all of his life, including during the Vietnam War. McCloy had an essential role in selecting targets for non-destruction, which meant other targets were selected for destruction.

Another individual involved in the Air Force command and target selection was Trubee Davison, who also had close contacts on Wall Street. Davison had been the assistant secretary of war for air between the wars. However, Davison's first association with the air force was during his years at Yale during WWI. At Yale, Trubee formed the special Yale Unit of the Naval Reserve Flying Corps. The unit was closely associated with Skull and Bones.

So, the Jesuit's Illuminati organization of Skull and Bones had members who not only were big Wall Street players but also played a large role in target selection in the air war over Germany. These men had links to the same corporations that had subsidiaries in Germany and these subsidiaries were removed by them from the list of targets given the Allied bombers.

A more thorough investigation of factories was made later in March. The American liberators were stunned that the German Ford plant was nearly intact. The only damage the plant received was from artillery shells during the pointless last stand by the Nazis. The plant was not in operation as much of the equipment had been removed and shipped across the Rhine to hidden locations.

The able wartime director of Ford Works, Robert Hans Schmidt suggested to the allies that he could produce 500 trucks in a short time, if he was allowed to conserve material, which was available. On May 8, 1945, the day after Germany's unconditional surrender, an American documentary team set up its cameras to record the sight of the first postwar truck coming off the Ford Werke assembly line.

The investigation of Ford Werke would not begin until June a month after Germany's surrender and just before the Cologne area was passed over to British control. The investigation was headed by Henry Schneider. On September 5, 1945 Schneider presented his report entitled Report on Ford Werke Aktiengesellschaft to General Clay. Schneider's team did not learn about a further set of meaningful documents kept by Johannnes Krohn, the Reichskommissar for the Treatment of Enemy Property. Nevertheless, the conclusions they made proved highly reliable. An excerpt from the report follows:

"The Reich used German Ford and its cooperative parent in Dearborn as a direct means of stockpiling the raw materials needed for war. Even prior to the War, German Ford arranged to produce for the Reich vehicles of a strictly military nature This was done with the knowledge and approval of Dearborn.

When war came, German Ford stepped into the position of a major supplier of vehicles for the Wehrmacht. In addition, as much as 7% or 8% of total output during the war years consisted of more specialized war material.

As was common in other German enterprises, Ford increasingly resorted to use of prisoners of war and other slave labor who had to live behind barbed wire. The foreign prisoners employed rose to over 40% of its labor supply in 1944..."

Despite the report on Ford Werke and the Senate investigations, no charges would ever be filed against Ford for trading with the enemy or using slave labor. Ford Werke was now rehabilitated; it was producing trucks for the revival of the German economy as a bulkhead against communism. The deaths of thousands of GIs on the battlefields of Europe were in vain. Betrayed by an elite cadre of their own countrymen, the war against fascism was lost. Under a cloak of free enterprise and anti-communism American fascist were now taking control.

Throughout Europe and in Germany in particular, the scene was much the same. Large industrial plants stood unscathed amid a field of rubble especially those plants that had connections to American firms as the Ford and I.G. Farben plants at Cologne. In fact, the I.G. Farben building in Berlin was untouched and used by the allies as a command center. It stood in stark contrast to the rest of the city, which lay in ruins.

Hans Kammler was ordered by Hitler to execute the Peenemunde scientists to stop them falling into Anglo-American hands.

It was a huge risk to Kammler's own personal safety to evacuate all the Peenemunde staff in April 1945 to Oberammergau, Bavaria, contrary to Hitler's orders. Kammler did so to create a bargaining chip for negotiations with the OSS at the war's end.

Some of the scientists that Kammler saved were Werner von Braun and Dr. Debus who were later brought to the U.S. in Operation Paperclip, along with thousands of others. Even later, the two would ended up directing NASA. Operation Paperclip basically cleaned up records of Nazi scientists and technical experts to bring them into the United States after the U.S. State Department forbade issuing visas to any former Nazis.

Kammler himself was charged, in his absence, for war crimes at Nuremburg because he was in charge of Nordhousen where slave labor was used under terrible conditions, resulting in much unnecessary death. So, when he was brought to Wright Patterson Field, he had to undergo an identity change.

Kammler would also later end up at NASA, working under a different identity. He was recognized by NASA employee, Clark C. McClellan in an office meeting at NASA with Dr. Debus.

McClellan became good friends with both Dr. Wernher von Braun and Dr. Debus and learned much from both of them that has been kept secret from the public. Being a patriotic American, McClellan believed that the public had a right to know and started revealing what he discovered from these former Nazi scientists at his website: http://www.stargate-chronicles.com

Also, much of McClellan's amazing discoveries are described in my book, *Inside the Secret Space Programs.*

NASA decided to punish McClellan for doing this by stripping all his benefits and pension from working at NASA. Now, McClellan is barely surviving from his social security.

Because of the intelligence operation of Admiral Rico Botta, U.S. Generals like Patton were informed just where to look for what was left of advanced German technology after the defeat of Germany. Patton was diverted from his race to beat the Russians to Berlin in order to recover this German secret advanced technology.

Patton located Hans Kammler and Die Glock (or the Nazi Bell) and had them both sent back to Wright Patterson for analysis and back engineering. Later, Patton was killed to prevent him talking about what he knew.

Naval intelligence also knew just where to find the secret New Berlin base in Antarctica. After World War II, in 1945, when U.S. authorities thought that the militarily defeated Germans might be willing to cooperate, they organized a secret mission to the German base in Antarctica.

The pilot chosen for this secret mission was Naval Commander, Graham Bethune. Bethune had a top-secret security clearance, had graduated from Academy Air at Pensacola, Florida in 1943, and was an excellent celestial navigator, He also had a lot of night flying experience, hunting German Submarines during the war.

Naval Commander, Graham Bethune, flew Admiral Richard E. Byrd, along with British and American scientists to Antarctica to meet with German scientists as an attempt to have the Germans reveal the secrets of their flying saucers. This mission was a total failure as the Germans refused to reveal their secret technology. Commander Bethune then flew Admiral Byrd back to the Pentagon to report on this matter.

As a result of the lack of cooperation by the Antarctic Germans, who failed to go along with the official surrender of Germany, Operation Highjump was planned. Operation High Jump was presented as a scientific expedition but in reality was a military operation intended to destroy the secret German New Berlin base.

Operation High Jump, officially titled *The United States Navy Antarctic Developments Program, 1946–1947*, was a U.S. Navy operation organized by Rear Admiral Richard E. Byrd Jr., USN (Ret), Officer in Charge, Task Force 68, and led by Rear Admiral Richard H. Cruzen, USN, Commanding Officer, Task Force 68. Operation High Jump commenced 26 August 1946 and ended in late February 1947. Task Force 68 included 4,700 men, 13 ships, and 33 aircraft.

Highjump was a massive Naval operation which included an aircraft carrier, destroyers, a submarine, two seaplane tenders, two icebreakers and other support ships. Admirals Byrd and Cruzen led the operation, which split up into two groups which would converge on Neuschwabenland from the Weddell Sea. A third group went to the old Little America base and improved it, adding a long airstrip that DC-3s could use.

The early termination of the expedition was caused by the military losses incurred by this South Polar fleet.

On Jan 17, 1947, two days after Byrd's central force made the

Bay of Wales, strange lights were spotted in the distance in the early morning darkness, by radioman John P. Szelwech aboard the *USS Brownson*. They did not show up on radar. This was entered into the ship's log. About 3 hours later, the lights reappeared and rapidly approached the destroyer.

Commander Gimber ordered the ship's 40 mm Bofors anti-aircraft guns and 20 mm Oerlikon cannons to open fire on the rapidly approaching craft, which rapidly flew at about 200 feet above the *USS Brownson*. This anti-aircraft fire seemed to have no effect on these strange craft.

This encounter was the beginning of several brief but fierce squirmishes that lasted over the next several weeks, in which dozens of officers and men were killed. This led to the evacuation of Byrd's central command from the Bay of Wales on February 22 by the icebreaker *Burton Island*.

As a parting shot, the Germans attacked the retreating fleet again on February 26. This attack was witnessed by flying boat pilot, John Sayerson, who said the following:

> "The thing shot up out of the water at a tremendous velocity as if pursued by the devil and flew between the masts [of the ship] at such a high speed that the radio antenna oscillated back and forth in its turbulence. An aircraft from the *Currituck* that took off just a few moments later was struck by some kind of an unknown ray from the object, and almost instantly crashed into the sea near our vessel.
>
> I could hardly believe what I saw. The thing flew without making any sound as it passed close over our ships and harmlessly through their lethal anti-aircraft fire.

About 10 miles away, the torpedo boat *Maddox* burst into flames and began to sink. Despite the danger, rescue boats went to her aid before she sank 20 minutes later.

Having personally witnessed this attack by the object that flew out of the sea, all I can say was that it was frightening."

Officially, Operation High Jump's primary mission was to establish the Antarctic research base Little America IV. After eight weeks, Operation High Jump was defeated by German flying saucers and Byrd beat a hasty retreat back to the U.S.

Admiral Byrd made the following statement to a reporter for the Chilean newspaper *Brisant* while aboard the command ship, *Mount Olympus* on the voyage home:

"It was necessary for the USA to take defensive actions against enemy air fighters which come from the polar regions...Fighters that are able to fly from one pole to the other with incredible speed."

Also, Admiral Byrd, in an interview with Lee van Atta of International News Service aboard the expeditions command ship, discussed the lessons learned from the operation. The interview appeared in the Wednesday, March 5, 1947 edition of the Chilean Newspaper El Mercurio and read in part as follows:

Admiral Richard E. Byrd warned today that the United States should adopt measures of protection against the possibility of an invasion of the country by hostile planes coming from the polar regions.

The Admiral explained that he was not trying to scare anyone, but the cruel reality is that in case of a new war, the United States could be attacked by planes flying over one or both poles. This statement was made as part of a recapitulation of his own polar experience, in an exclusive interview with International News Service.

Talking about the recently completed expedition, Byrd said that the most important result of his observations and discoveries is the potential effect that they have in relation to the security of the United States. The fantastic speed with which the world is shrinking – recalled the Admiral – is one of the most important lessons learned during his recent Antarctic exploration.

I should warn my compatriots that the time has ended when we were able to take refuge in our isolation and rely on the certainty that the distances, the oceans, and the poles were a guarantee of safety.

A 2006, Russian documentary made public for the first time, a 1947 secret Soviet intelligence report commissioned by Joseph Stalin of the Operation Highjump mission to Antarctica. The intelligence report, gathered from Soviet spies embedded in the US, revealed that the US Navy had sent the military expedition to find and destroy a hidden Nazi base. On the way, they encountered a mysterious UFO force that attacked the military expedition destroying several ships and a significant number of planes. Indeed, Operation Highjump had suffered "many casualties" as stated in initial press reports from Chile. Here is the documentary: https://www.youtube.com/watch?v=MwUpPwyyvLw

Admiral Byrd arrived back in Washington D.C. on April 14, 1947 where he was debriefed by Naval Intelligence. Later, it was reported that Admiral Byrd flew into a rage while testifying before

President Truman and the Joint Chiefs of Staff, and strongly suggested that Antarctica be turned into a nuclear testing ground. Much of Operation High Jump is still classified today.

Other events, besides the German situation - such as the recovery of several extraterrestrial flying saucers before the famous Roswell incident - also increased the need for much more government secrecy to keep the people from panicking and insure stability. These events lead to the creation of the National Security Act of 1947 and the ultra-secret MJ-12 group to handle the extraterrestrial situation.

And, other events had taken priority over, far away, Antarctica, like the rebuilding of war torn Europe and Japan and the Korean war. But, in Korea, negotiations began for a ceasefire in July 1951 at Panmunjom.

So, five years after the failure of Operation High Jump to destroy New Berlin, President Truman was seriously considering destroying the Nazi Antarctic base with a nuclear bomb. Many of the Paperclip Nazis brought into the U.S. after the War and hired by U.S. corporations still retained their loyalty to the, now secret Nazi cause or the Nazi International that researcher, Joseph Ferrell has uncovered and documented in his book, *The Nazi International.*

These Paperclip Nazis got wind of Truman's plan to use an atomic bomb on the Antarctic base and informed their compatriots down there via Argentina. This led to several massive overflights of Nazi UFOs, launched from Antarctica, over Washington D.C. in 1952.

Fighter jets were scrambled to intercept these UFOs but they were totally out maneuvered and helpless to defend the Capitol!

This show of force by the Nazi International let Truman know that the U.S. Capitol could also be destroyed with Nazi superior technology, which by this time included nuclear weapons. This led to secret negotiations between the Truman Administration and the secret Nazi International, using some Paperclip Nazis as intermediaries.

Truman did not like or agree with all the Nazi International demands, but did agree to leave their Antarctic base in peace. The negotiations continued into the Eisenhower Administration.

One interesting meeting between the Nazi International and Eisenhower took place at Holloman AFB, near White Sands Proving Ground in New Mexico, on February 11, 1955, according to NASA employee Clark McClellan, who got the information from another NASA employee and friend, Ernst Steinhoff.

Steinhoff had worked at Peenemunde in charge of the German V-2 Program and submarine launched missiles. After the war, he was brought to the U.S. in operation Paperclip and in 1949 worked at the White Sands Proving Ground and nearby Holloman Air Force Base.

On February 11, 1955, Air Force One, a Lockheed Constellation, named *Columbine III* and carrying the U.S. President, Eisenhower, landed at Holloman AFB at 9:00 AM. The plane stopped in the middle of the runway and parked there. All Radars were ordered to be turned off. Soon, two flying saucers were visually spotted approaching the airbase.

One saucer landed opposite Air Force One, while the other remained in the air, observing events below. Eisenhower disembarked from the *Columbine III* and walked over to the landed saucer and entered it, according to witnesses of the event. 45

minutes later, he departed from the saucer and returned to his plane which then proceed to the terminal. The President stayed at Holloman until about 4:30 or 5:00 PM visiting with Air Force personnel, and then left. (8)

According to what McClellan was told by Dr. Ernst Steinhoff, who was present at Holloman at that time, the saucers were the German V-7s, which the Nazis had improved since they were first operational in 1942 at Peenemunde and it was German officers – not extraterrestrials, as some have supposed – that Eisenhower was meeting with!

In a later meeting on February 1960, President Eisenhower had traveled to San Carlos de San Bariloche, Argentina, where officially he negotiated the Joint Declaration of Bariloche with Argentinian President, Frondizi concerning Peace and Freedom in the Americas.

However, San Carlos de San Bariloche, at the time, was the secret headquarters of the Nazi International. Adolf Hitler, and his wife, Eva lived nearby at their Santa Clara ranch after having been moved there from their Inalco estate in 1955. By then, Hitler's health was failing and Martin Bormann was actually in charge. The real topic of negotiations concerned deals which would further put the U.S. Military Industrial Complex firmly under the control of the Nazi International.

Finally, after much negotiation, a secret treaty was agreed to. This treaty basically handed over to the Nazis, the U.S. Military Industrial Complex. In return, certain U.S. corporations would be allowed access to their Moon base and their secret Antarctic base.

This basic sellout to the Nazis by Eisenhower was supposed to be a stalling action until the U.S. could develop equivalent technology and have a fighting chance against the Nazis.

But, the Nazis were technically far ahead of the U.S. And, these Nazis were also taking over the CIA, FBI and NASA. Also, the same U.S. industrialist that supported fascism before World War II and the Jesuit's secret societies like the Illuminati and Skull and Bones helped in this Nazi takeover of the U.S. The powerful Bush and Harriman Families, who had several members in Skull and Bones also both played a large role in this takeover.

By 1960, the silent and secret victory of the Nazis over the U.S. was complete. Eisenhower tried to warn the American people about the Military Industrial Complex in his farewell speech. But, the people just didn't understand the danger from this hidden direction. And, government secrecy did not help matters either! (9)

Otto Skorzeny, the favorite commando of Adolf Hitler, was inducted into the CIA while still working for the Nazi International and ODESSA. In his death bed confessions, revealed in the book *The Bush Connection* by Eric Orion, Skorzeny reveals that he worked with George H.W. Bush, who joined the CIA while at Yale as well as being in Skull and Bones. The two worked together to place many former Nazis into the CIA just after its formation.

Skorzeny claimed that George H.W. Bush and his father, Prescott Bush were not really who they claimed to be. Prescott Bush was really the alias of George H. Scherf. And, George H.W Bush was really George H. Scherf Jr.

George H. Scherf was really a German born spy who got a job as Nikola Tesla's trusted assistant. He also used his Bush alias to work as a director of Union Bank in New York, which helped bankroll The Third Reich and also become a Senator of New York.

Don Nicoloff published an extremely thorough and well researched treatise in two parts in April 2007 in the Idaho Observer titled "Deathbed Confessions, photos support claims that George H. Scherf(f), Jr., was the 41st U.S. president"

Don Nicoloff takes the Skorzeny information to a much higher level of investigative research. He had already been researching the Bush family genealogy before learning about Skorzeny's death bed confession. This confession helped him put together many loose ends and verify the Skorzeny information. The Skorzeny photo at the top of Don Nicoloff's article is interesting. It shows a "family" group pose likely taken in a Bavarian or Alpine lodge location owned by the Scherff family in the 1930's that purports to include Martin Bormann, Reinhard Gehlen, Josef Mengele, Otto Skorzeny, George H. Scherff, Jr (GHW Bush), Prescott Bush, "Mother" Scherff, and likely Dorothy Walker Bush. This photo can also be seen here: http://educate-yourself. org/cn/familythatpreystogethercompared24aug07.shtml

Another site that is doing extensive research on the dubious genealogy of the Bush Family is here:

http://www.bibliotecapleyades.net/sociopolitica/esp_sociopol_ bush19.htm

Other revelations by Otto Skorzeny include:

Skorzeny's role with Reinhard Gehlen in the assassination of Nikola Tesla by suffocation on January 6, 1943 after tricking Tesla, on January 5, into revealing the full details of his most important discoveries and then stealing his most valuable inventions, along with the contents of his safe, which were then delivered to Hitler.

The faked April 1945 "suicide" of Hitler in the Berlin bunker, corroborating the 1948 CIA interrogation of Gestapo Chief Heinrich Mueller, the Nordon Report, and the Russian 1946 Investigative Committee report to Stalin.

Skorzeny's pivotal role in consolidating the intelligence assets of the Nazi Gestapo and the Nazi-infiltrated OSS into the CIA in 1947, working with George H.W. Bush, Nazi SS spy master Reinhard Gehlen, "Wild Bill" Donovan and Allen Foster Dulles.

Skorzeny claimed central responsibility for setting up the ODESSA operation to surreptitiously relocate 50,000 Nazis into North and South America with CIA complicity and protection. Later, the CIA's Operation Condor would use many of these relocated Nazis.

Skorzeny claimed that Reinhard Gehlen, Josef Mengele, and George H.W. Bush directly participated in JFK's assassination.

These revelations, if proven correct, demonstrate just how little the American people know what is really going on around them. It also shows that the CIA has been basically a Nazi operation or at least worked closely with the Nazis and kept it all hushed up since the end of World War II.

Other Independent evidence to bolster this conclusion was the fact that the CIA actually hired Knight of Malta, Reinhard Gehlen and his entire intelligence operation. Reinhard Gehlen was the Nazi general who was chief of the Foreign Armies East military intelligence operations for the Third Reich. After post war Germany regained its sovereignty, Reinhard Gehlen became the first president of West Germany's BND intelligence service in 1956.

While, at the CIA, Gehlen was feeding disinformation to the President and the Pentagon. This was part of the Nazi International policy of causing the U.S. to forget about the Nazis and focus on the "Russian Threat" and in effect to get the U.S. to continue the Nazi's war against our former World War II Allies – the Soviet Union!

This disinformation, maintained that the Soviets were getting ready to invade Western Europe. The Soviets had no such plan. They were too busy trying to rebuild their war-torn countries to even consider such a foolish military adventure. They had lost too many millions of their best fighters in their battle with the Nazis and were not about to take on the Americans and British, who were guarding Western Europe at the time.

However, many heeded Gehlen's disinformation, especially those in the weapons industries who were worried about losing their lucrative government contracts during peace time. And of course, the Jesuits, who had invested heavily in many of these same defense companies, also weighed in.

Since, the Jesuits also controlled the CIA, Gehlen's disinformation was touted as reliable. Thus, began the steps leading to the "Cold War".

Another person that influenced U.S. entry into the Cold Car was William Averell Harriman.

Beginning in the spring of 1941, Harriman served President Franklin D. Roosevelt as a special envoy to Europe and helped coordinate the Lend-Lease program with Briton. He was later dispatched to Moscow to negotiate the terms of the Lend-Lease agreement with the Soviet Union. At first, the American public was skeptical of a lend lease with the USSR, but became more favorable to the idea after the Japanese attack on Pearl harbor.

Interestingly, before the U.S. entry into World War II and even some time after, the Harriman interests were actively assisting the Third Reich! Harriman's banking business was the main connection for German companies and the varied U.S. financial interests of Fritz Thyssen who was a financial backer of the Nazis until 1938.

The Trading with The Enemy Act classified any business transactions for profit with enemy nations as illegal, and any funds or assets involved were subject to seizure by the U.S. government. The declaration of war on the U.S. by Hitler led to the U.S. government order on October 20, 1942 to seize German interests in the U.S. which included Harriman's operations in New York City.

The Harriman business interests seized under the act in October and November 1942 included; The Union Banking Corporation, The Holland-American Trading Corporation, The Seamless Steel Equipment Corporation and the Silesian-American Corporation.

The assets were held by the government for the duration of the war, then returned afterward. However, Harriman, himself was never punished. Harriman served as ambassador to the Soviet Union until January 1946. The Lend Lease Program with the USSR was terminated immediately after the surrender of Germany. Ships carrying cargo to the USSR were ordered to return.

When Harriman returned to the United States, he worked hard to get George Kennan's Long Telegram into wide distribution. Kennan had spent a long time in Russia and was considered an expert on Russian affairs. Kennan's analysis, which generally lined up with Harriman's, became the cornerstone of Truman's Cold War strategy of containment of the USSR.

However, in later years, Kennan would consider the U.S. policy of Soviet containment as being way too extreme. Kennan argued that containment did not demand a militarized U.S. foreign policy. Kennan was opposed to the U.S. Vietnam policy and later, the U.S. destruction of Yugoslavia and the expansion of NATO by the William Clinton Administration.

So, the Nazi business partner, William Averell Harriman continued to help the Nazi cause by helping the Nazis get the U.S. to continue their war against the USSR – our former ally in the war against the Nazis! A close partner of Harriman was Prescott Bush, another director of Union Banking Company and father of President George H.W. Bush. All three men were members of the Illuminati secret society of Skull and Bones and were ardent Nazi supporters.

The Cold War ended in 1991 with the collapse of the Soviet Union. However, the chain of events leading to this unexpected collapse can be traced to the signing of the Helsinki Accords during the Gerald Ford Administration.

Before the signing of these Accords by the U.S and U.S.S.R., there existed no legal framework for human rights within the Soviet Union. Diplomacy between the superpowers always recognized that outside countries were not to interfere with the domestic policies of other nations and only foreign policies, which usually centered around national security issues, would be addressed in this diplomacy. And the Soviets focus was on productivity and national security – not human rights.

The Helsinki Accords also made certain human rights a part of the accords. This gave a legal standing for human rights activists within the U.S.S.R. to make their case to the central communist party. And this started an ever-growing movement within the Soviet Empire for more freedom. This ended up with many communist countries withdrawing from the Soviet Union. When Russia herself withdrew from this union, it was the end of the old Soviet Union.

The present Russia is no longer communistic with state owned commercial enterprises. Russia recognized that more incentive was required to increase productivity. And this incentive was supplied by privatization.

At first, criminal oligarchs would buy up state owned companies and asset strip them, while placing the proceeds in secret numbered accounts in Swiss banks. Russia was quickly headed to becoming a failed state. But, Vladimir Putin soon put a stop to this process and criminalized this practice. Many of these criminals fled Russia for Israel and the United States where they obtained asylum. Under Putin, the privatization process was stabilized. And Russia has gradually rebuilt it's economic and military strength since the 1990s.

The government owned housing, which was formerly rented out to the tenants, was simply given to the tenants. These privately-owned properties could then be bought and sold like real estate in the U.S. There is presently a large amount of new housing being constructed there. But, the new houses have to be paid for. This has driven housing costs up in Russia - like in the U.S.

Many Russian companies, but not all, are also privately owned and operated. The ones that aren't are considered vital to Russian national interests, like energy and weapons development companies. Some companies are jointly owned both privately and by the Russian state.

I recently visited Saint Petersburg, Russia and found the people there very happy, friendly and proud to explain that Russia is no longer a communist country. The economy there was doing as well or better than most other Northern European countries I have visited, like Iceland, Sweden, Denmark and Germany. This is in spite of Anti-Russian sanctions and lower oil prices. In fact, Russia claims that the Sanctions actually helped Russia to become more independent and self-reliant.

Another change from the old Soviet Union is the re-emergence of the Eastern Orthodox Christian religion in Russia, which now stresses religious freedom along with freedom of speech and press. Russia is democratic and regularly holds elections.

So, the Russian Federation under Vladimir Putin is nothing like the old Soviet Union, under Joseph Stalin. And, the monopolistic dreams of the men who financed the Bolshevik Revolution are now falling apart.

Neuschwabenland and Nue Berlin

While the German war machine was being restored and continuing into World War II, there was a large, simultaneous project being carried out by Germany in the far away continent of Antarctica.

Hitler had entered into a treaty agreement with the Draco Reptilians in the 1930s. The meetings between Hitler and the Reptilians were set up by the psychic, Karl Haushofer of the secretive Swartz Sonne or Black Sun Society.

Karl Haushofer, had been initiated into the Black Dragon Society while in Japan instructing the Japanese military in 1909. At the inner core of the Black Dragon Society was another society - the Green Dragon Society comprised of Buddhist Monks that communicated with the Buddhist "Green Men" living in communities in Tibet.

This communication over the great distance between Japan and Tibet took place on the astral level where distance posed no problem. These Green Men got their name because they were communicating psychically with other men, covered with green scales, living underground - who were the Draco Reptilians!

Karl Haushofer had been one of the few from the West initiated into the Green Dragon inner circle of the terrible Black Dragon Society. There, he was psychically trained to communicate with the Green Men and the Reptilians. Part of his training also gave him the ability to see into the future. And, he retained these abilities on returning to Germany in 1911.

During World War I, Haushofer became a war hero because of his uncanny ability to predict where the Allies would strike next and come up with a counter strategy. In 1918, Haushofer and Rudolf von Sebottendorf founded the Thule Society in Munich, Germany.

Later, while Adolf Hitler was in prison, he would receive coaching from Haushofer in the ways of enhancing his own physic powers.

In any case, directions were obtained psychically by Haushofer on where and when Hitler and the Reptilian leader could meet in an underground cave. There, Hitler and his military guards met face to face with actual Reptilians which frightened him and his crew considerably. He referred to the Reptilian leader as a superman with terrible eyes. Hitler was able to overcome his fear in order to accomplish what he considered a greater goal – the military return of German greatness after the humiliations of Versailles following World War I.

More of these meetings followed, in which various provisions were agreed upon. These secret treaty agreements were not committed to paper. They were verbal agreements which were, nevertheless, actually kept as far as each side was able to do. In this treaty, Hitler agreed to use Germany's military power to help the Draco Reptilians conquer the Earth in an alliance with them.

In return, these Reptilians offered Hitler advanced technology and weaponry, an abandoned Reptilian underground city in Antarctica that could be reached by their U-boats, and the right to create a base in their sector on the far side of the Moon.

Later, many of the Green Men Buddhist from Tibet were brought to Germany by the Nazi SS. The reason for this has always posed a mystery to most historians. But, now we may get a hint of what their purpose may have been - especially after Hitler later turned on Haushofer and had him arrested and imprisoned and his talents could no longer be used!

These agreements between Hitler and the Reptilian leader included directions to one of their abandoned cities, located underground in Antarctica. Above this underground city existed geothermal springs that warmed the surface and created ice free zones and unfrozen lakes above.

These geothermal sources of heat also created warm rivers under the ice cap that melted tunnels in the ice that allowed submarines to approach this underground city which existed at the water's edge, under the icecap.

The seaplane carrier *Schwabenland* was used for the German 1938 Antarctic expedition, commanded by Captain Albert Richter, a veteran of cold weather operations. The ship carried several Dornier sea planes on board. These planes could be launched by catapults from the ship and later land on the water near the ship on floats and be reloaded back on board via onboard cranes.

Spiked steel poles with German swastika flags were loaded on board the seaplanes and dropped at spaced intervals as the seaplanes flew a course that marked the borders of

Nueschwabenland – the Antarctic land claimed by the Third Reich. This land contained about 600,000 square kilometers from 10 degrees West to 20 degrees East.

At the same time, German U-boats, following the instructions given the Germans by the Reptilians, and working their way underwater through the warm water, melted tunnels in the ice cap, found the huge cave and abandoned Reptilian city within. The city was in considerable need of rebuilding and repair.

After this discovery, more U-boats would bring workers and equipment to refurbish this city and build U-boat pens at the water's edge, deep under the ice cap. This underground city was named Nue Berlin. (New Berlin) and this refurbishing continued and even increased all throughout World War II.

After the German invasion of Ukraine, many Ukrainians were sent there to work as slave labor. Ten thousand, racially pure, Ukrainian, blond and blue eyed, young women were selected by Himmler, who was in charge of this secret project, as Antarktisches Siedlungensfrauen (Antarctic Settlement Women) and sent to Antarctica by Submarine. Himmler also selected 2,500 battle hardened Waffen-SS soldiers from the Eastern front to be sent to Nue Berlin to mate with these ladies and start to populate Nue Berlin with racially pure stock.

Newer and larger, XX1 U-boats, with radar proof snorkels, were designed to specifically to carry cargo for long distances, completely underwater for this project.

Near the end of the war, many of the German flying saucers were also sent to Neu Berlin. These Saucers had the capability to operate underwater, in the atmosphere and some even in outer space. So, arriving at and departing from Nue Berlin was no problem for these saucers and special hangars were constructed for them in the underground city.

Also at this time, according to researcher Rob Arndt, the entire Vrill and Thule Societies disappeared along with the entire SS technical branch! This disappearance also included 54 U-boats, 6,000 scientists, 142,000 to 250,000 people and tens of thousands of slave laborers. Were these people also sent to Nue Berlin? It is quite likely.

All of this activity, even with the long-distance submarines, did attract the attention the British, who also had outposts near Antarctica in the Southern Atlantic, like the Falkland, and South Georgia Islands.

In Fact, British Intelligence had known about the German underground base since 1939. But they only possessed scant knowledge and had no idea how big of an operation Nue Berlin actually was.

To learn more about this hidden German base, the British constructed their secret Maudhiem base, which was the largest of several secret bases Britain possessed in and around Antarctica.

The British had set up these bases after learning about the German base with the idea that there would be an eventual confrontation with the Germans in Antarctica. Maudheim was about 200 miles from the Muhlig-Hoffmann mountain and was about this distance from Neu Berlin which lay hidden deep below.

By October 1945, in Operation Taberin, 10 highly trained British SAS Commandos launched an attack through a tunnel in the Muhlig-Hoffmann mountains which led to Nue Berlin. This tunnel had been discovered and explored the previous year. These men were loaded with high explosives and were planning to sabotage the base They obviously had no idea how big of an operation they were up against.

When they actually discovered the city, they were amazed. People were crawling all over the place, which was built up so much that it looked like the Germans had been there for many decades – not just 6 or 7 years. They saw the sub pens and hangars for strange looking craft, that they had no idea how they could fly in or out of the underground base

There were well armed German military guards everywhere and the Commandos stayed hidden in the shadows. They eventually spotted an ammo dump and decided to place charges to blow it up. They also spotted places to place charges that would create rock falls that would cause a lot of damage.

Before they could accomplish these tasks, they were spotted by the German guards. Soon, these guards were chasing these Commandos through the tunnel they had entered by. A lot of gunfire was exchanged and only three of the Commandos made it back to the tunnel entrance, which they blew up - totally blocking it to the Germans.

Then, they radioed for an evacuation and were flown to South Georgia Island. There they were debriefed and told that this operation was to remain totally secret. It is thought that British shared intelligence on Operation Taberin with U.S. Intelligence, because it would take a much larger operation to take out the German base than the British originally thought, or could handle alone.

Hitler's Escape to Argentina

Many historians maintain that Hitler and his mate, Eva Braun, committed suicide in his bunker in Berlin just before the Soviets took that city. Painstaking research by other historical groups, like the Shark Hunters, and researchers like Simon Dustan and Gerrard Williams in their book *Grey Wolf,* came to a different conclusion.

Martin Bormann was the primary architect of Plan "B" after it was realized that Germany would lose the war in 1943. Plan B had a number of sub programs like Aktion Feuerland (Project Land of Fire) and Aktion Alderflug (Project Eagle Flight).

Aktion Feuerland was a plan to create a safe refuge for Hitler in Argentina. There already existed a number of German settlements in Argentina, which was generally friendly to the Third Reich. Hundreds of square kilometers of land had been brought up by Germans around Lake Nahuel Huapi and the German settlement of San Carlos de Bariloche in Argentina. This location was close to the Andes and had a climate and appearance similar to Hitler's favorite place – the Berghof in Bavaria. A place had already been built for Hitler in 1943 called Inalco on the shore of the lake.

Aktion Alderflug was a plan to redistribute German wealth into foreign banks while disguising the actual ownership of these accounts. Between 1943 and 1945, more than 200 German companies set up subsidiaries in Argentina. These companies would later provide employment to the many thousands of Nazis fleeing Europe to Argentina after the war. Also shares of stock equities were brought up by Bormann in foreign companies, primarily in North America. In this project, Bormann took lessons from the master of this sort of thing – I.G. Farben - which had acquired a voting majority in 170 American companies and minority holdings in another 108 by the time the United States declared war on Germany in 1941!

By the time the European war broke out in 1939, there were already 237,000 Germans living in Argentina, 60, 000 of which were members of the Nazi party. Also, the Argentinians largely supported Germany because of their fear of British encroachment of their territory. In World War I, Briton had attempted to claim all of South America South of the latitude of the Falkland Islands, which Argentina also claimed.

All of these factors made Argentina the ideal location to hide Hitler and many thousands of other Nazis fleeing Europe. Bormann would also be the man in charge of Hitler's escape just before the end of the Third Reich.

The new Reich Chancellery was built to be the seat of power for Hitler's Thousand-Year Reich. However, when Allied Bombing of Berlin intensified, The Fuhrer Bunker was constructed by this Chancellery to protect the leadership of this center of Nazi power.

The last two months before the end of Germany's war, scenes in the Fuhrer Bunker were pretty dismal. Not only were the underground quarters quite cramped, but the continual sound of a diesel generator running nearby 24/7 was subtly annoying. Russian artillery rounds exploding could be heard in the distance while Allied bombs could be actually felt in the bunker.

Even under 23 feet of soil and 11 feet, three inches of reinforced concrete, one had to steady their cup of coffee with two hands to keep it from being spilt when an Allied bomb would go off nearby. Some of these bombs would break water pipes on the streets overhead and cause water leaks into the bunker.

Intelligence coming in to the bunker from German military forces was invariably bad news, which would often throw Hitler into a violent rage that would frighten even his most seasoned intelligence officers and military veterans. Most thought that it was only a matter of time before they would all be dead.

On April, 15, 1945 Eva Braun came to the bunker from Bavaria. On April 16, all stored records were ordered burned and new reports were burned each day after being read by Hitler. On April 21, Martin Bormann and Admiral Donitz appeared at the Fuhrer Bunker. The personnel there were told that they would soon be moving somewhere else to carry on their work. They took the news with great relief but many wondered if they would even escape Berlin alive.

On April 28, 1945, Hitler, his dog Blondi and Eva were taken from the Bunker with a group including; SS General Hermann Fegelein, trusted husband of Eva Braun's sister Gretl, and 6 trusted soldiers of SS Guard Battalion.

Hitler and Eva both seemed drugged and listlessly followed the SS officers out. A short time later, a double of Hitler and Eva were lead into the bunker according to eye witnesses, including Nazi intelligence officer, Angel Alcazar de Velasco. In fact, Hitler had been forcibly drugged on the orders of Martin Bormann.

It was obvious that Martin Bormann was the man really in charge at this time. Previously, Hitler had said that that he preferred to die fighting the Russians with his men in Berlin than flee as a coward. Bormann felt the importance of preserving Hitler as a Nazi leader to keep Nazi cause alive. His mission was to keep Hitler alive. To do that, he had to get Hitler out of Berlin and to a safe refuge, even if it meant drugging Hitler to make him submissive and obedient.

Hitler's group passed through a secret tunnel system connecting with the U-Bahn and walked 4 miles through the underground U-Bahn system from the Fuhrerbunker to Fehrbelliner Platz. There, they met with Eva Braun's sister Ilse and Fegelein's close friend SS General Joachim Roemer. Awaiting the group were 3 Tiger II tanks and an armored personnel carrier to carry them half a mile to an airstrip at Honenzollerndamm. There, they boarded an awaiting JU-52 aircraft.

A Ju-52 piloted by Peter Erich Baumgart then flew the party (minus the six soldiers of the SS Guard Battalion), on a zigzag flight path to avoid Allied bombers and fighters, to Tonder in Denmark. There, the party disembarked. Hither shook Baumgart's hand and then slipped a check, signed by Hitler, for 20,000 Reich marks into Baumgart's hand.

The Party then boarded another JU-52 and then flew to Travemunde on the German coast. There was a JU-252 waiting for them with engines already running. Now, the party split up, with General Joachim Roemer and his wife, SS General Hermann Fegelein's wife Gretl Braun, and Ilse Braun (both pregnant, felt it unsafe to travel any further) all decided to remain in

Germany. Being in the western part of Germany, they wouldn't risk capture by the Russians, which they primarily feared.

Those remaining of the party boarded the JU-252 and were flown to Reus Military Base in Spain. The JU-252 then was ordered disassembled by order of Spain's leader, Franco, to hide evidence of Franco's complicity in Hitler's escape.

The next leg of the fugitive's journey would be a flight to Fuerteventura Island in the Canaries Islands, where they arrived on April 30. A secret U-boat refueling base had been created on the western end to the island and Hitler and his companions would stay at the nearby Villa Winter.

Meanwhile back in Berlin, on April 30, 1945, Bormann ordered Heinrich Muller to execute doubles of Hitler and Eva in the Fuhrerbunker. Then, Bormann communicated the "news" of Hitler's death to Admiral Karl Donitz, appointed as the new Reich president in Hitler's will. Donitz would later order General Jodel to unconditionally surrender to the Allies on May 7.

Goebbels and his wife and children, also at the Fuhrerbunker, committed suicide by poison, knowing that the Russians were only days from capturing all of Berlin and preferring death to what might happen at the hands of the Russians. All of these bodies were ordered burned in the Chancellery garden by Bormann.

The remaining officers left the Fuhrerbunker on May1 through May 2. On May 2, Martin Bormann; Warner Naumann, Goebbels successor; Artur Axmann, leader of the Hitler Youth; Ludwig Stumpfegger, Hitler's doctor; and Waffen-SS Captain, Joachim Tiburtius made their way out of the Fuhrerbunker and climbed aboard two Tiger II tanks. Traveling down a narrow street, the forward tank took a direct hit from a Soviet antitank weapon, blocking the other tank's path. Bormann and Tiburtius left the other tank and made their way on foot to the Hotel Atlas, where Bormann had previously stashed false identity papers clothing and money. Eventually, traveling in and out of enemy lines, Bormann would make his way south to the Bavarian Alps.

On Fuerteventura Island in the Canaries, Hitler and his companions left the Villa Winter, where they had been resting after their long flight. They were driven over dirt roads in the darkness of night to a pier on the northwestern point and there boarded a fishing boat which took them to Submarine U-518, a type IX long range U-boat, waiting nearby off shore.

U-518, commanded by Hans Werner Offermann, was one of three type IX U-boats sent to this location by sealed orders previously sent from Bormann himself weeks earlier. The other two were U-880, commanded by Gerhard Schotzau and U-1235 commanded by Franz Barsch. General Hermann Fegelein was taken to U-880. Precious booty was loaded aboard U-1235 and U-880 to be taken to Argentina as an ongoing part of Aktion Feuerland.

Hitler and Eva had to endure almost two months, mainly under water, in the cramped quarters aboard the fighting submarine U-158, as it transported them to Mar de Plata, Argentina. Hitler's dog Blondi quickly made friends with the Submarine's crew.

U-880, with SS General Hermann Fegelein, aboard, was traveling at top speed and arrived five days ahead of the Fuhrer's submarine. This would allow Fegelein time to make preparations for the Fuhrer's arrival.

The U-880's arrival off of Mar de Plata was on the night of July 22, 1945. A tugboat from the Delfino SA line met the U-880 about 30 miles offshore of Mar del Plata. Sailors from the U-880 offloaded forty small, ammunition sized, but heavy, boxes from the submarine to the tugboat.

Then, all but one of the U-880's crew and passengers also boarded the tugboat, while the remaining crewmember opened the submarine's sea cocks to sink the U-880 in the depths of the South Atlantic. Before the Submarine could sink, this crewman also boarded the Tugboat.

General Fegelein met Rodolfo Freude, Son of Ludwig Freude, who was the main Nazi representative in Argenta, and Colonial Juan Peron's personal representative, after boarding the Tugboat. Then, Fegelein took a shower and dressed up in a finely tailored business suit given him by Freude. In the wheelhouse, Freude and Fegelein met with the U-880's other passenger who had also just showered and dressed up, Willi Koehn, former head of Chili's Nazi Party and chief of the Latin American division of the German Foreign ministry.

When the Tugboat arrived at the quay of Mar del Plata, an Argentine Navy staff car awaited them. Fegelein and Freude entered the car and were whisked away. Later, they were flown in an Argentinian Air Force Curtiss Condor Biplane to a grass airstrip on the Lahunsen owned, Estancia Moromar Ranch four miles from the coast near Necochea. The Lahunsen conglomerate of companies was a known center for German espionage in Argentina.

In the cold night of July 28, at 1:00 AM, General Fegelein waited on the beach of Necochea, for the arrival of his sister-in-law and Fuhrer aboard the U-518. He was accompanied by a number of well-armed soldiers. The submarine arrived an hour later, cautiously approaching the beach as close as safely possible without running aground. Soon, a small shore boat approached the submarine and Hitler, Eva and Hitler's dog Blondi were assisted aboard the small boat.

After powering the small boat onto the beach through a surprisingly small surf for the South Atlantic winter, it's passengers were assisted ashore by General Fegelein's newly acquired soldiers (actually sailors from the *Admiral Gaf Spee* which had been scuttled off of Montevideo in December, 1939). The landing party and General walked from the beach to a waiting car which took them to the Estancia Moromar.

Meanwhile, the small boat and a number of rubber dinghies from the U-518 were off-loading more heavy boxes, similar to the ones off-loaded from U-880, from the submarine to trucks waiting on the beach. When this job was finished and most of the crew was ashore, a small crew took the U-518 further off shore, followed by the small boat.

After scuttling the U-518 in deeper waters, the small crew was returned to the beach in the small boat and rejoined their comrades. All the crew members, now dressed in civilian clothes, and the submarine cargo were taken to Estancia Moromar.

Although no mention is made of U-1235 in this account from the book *Grey Wolf*, it is quite likely that U-1235 also unloaded its cargo at about this time and place and was similarly scuttled in deep water. I base this premise on testimony given later by three sailors from the *Admiral Gaf Spee* at a 1952 Congressional inquiry led by Silvano Santander into Nazi activity in Argentina.

The three witnesses were; Alfred Schultz, Walter Dettelmann and Willi Brennecke, who claimed that between July 23 and 29, 1945 they unloaded cargo from two submarines on the Argentine coast. They weren't sure of the exact location but all agreed that it was near a Lahunsen ranch. Here is a summary of their testimony:

> "Two U-boats unload a huge number of heavy crates which were then shuttled to the house on the estate in eight lorries. Eighty persons disembarked in rubber dinghies. It was alleged the cargo was 'some of the treasure of the SS-RSHA' and 'documentation pertaining to the technical science of the secret weaponry'."

The SS-RSHA was the SS security service, headed by Hienrich Himmler. And the heavy crates quite likely contained gold bars.

After, cleaning up and a night's rest at the Estancia Moromar, Hitler's party was flown by the same Curtiss Condor Biplane that General Fegelein arrived in, to San Carlos de Bariloche, making a short refueling stop at Neuquen. The plane landed at dusk on July 30, at the Estancia San Ramon's air field. San Ramon was owned by the family of Prince Stephan zu Schaumburg-Lippe.

In 1945, the Germans had complete control over the access to Carlos de San Bariloche, including the Estancia San Ramon. Hitler and Eva Braun would stay for nine months at San Ramon under different names. Their radio contact with Bormann, who was still on the move in Europe was infrequent. But, Bormann's organization in Argentina was busily preparing Hitler's permanent residence at Inalco, 56 miles from San Ramon.

Eva previously had a daughter by Hitler named Ursula. Her parents had last seen her on April 11, 1945 in Bavaria where

Ursula was secreted away. By this time, Ursula was 6 years old and Eva missed her dearly.

So, arraignments were made to have Ursula shipped first class from Spain to Buenos Aries. Bormann had already made sure that Ursula was well schooled in the Spanish language and was given a Spanish passport. In September of 1945, her Uncle, Hermann Fegelein met her at the dock in Buenos Aries and had her flown to Estancia San Ramon in the same Curtiss Condor Biplane that her parents had arrived in. The reunion was quite a happy affair.

By this time Eva was pregnant again. She had conceived in March and now it was beginning to show. So, Hitler thought it best to get married for the proper appearance and for the sake of the children. And conveniently, there was a small Catholic chapel at San Ramon. Hermann Fegelein would give his sister-in-law away at the wedding.

In December of 1945, Eva Hither gave birth to a baby boy. His parents decided to name him Adolf Hitler Jr. That was not the name used around the Estancia – but Adolpho with an alias surname corresponding to that used by Eva and Hitler. The staff and other visitors of Estancia San Ramon thought that they made a charming family.

In late March, 1946, the employees at Estancia San Ramon were told that the charming family staying there had perished in a terrible car accident and their bodies had been burned beyond recognition.

So, not only had Hitler and Eva died in the Fuhrerbunker and had been burned beyond recognition, but they had also died in a car crash near Estancia San Ramon and were badly burned beyond recognition. If anyone looking for Hitler actually made it as far as San Carlos de Bariloche, the trail would again go cold.

Meanwhile, Hitler and his family was secretly moved to a more permanent home at Inalco, which was now fully prepared for them.

Back in Europe, Bormann, who was hiding near this Bavarian redoubt, was communicating with Ludwig Freude through a portable T-43 encryption system and remotely controlling operations in Argentina. He was also using his connections at the Vatican to obtain a Spanish passport. The most important Vatican connection was Bishop Alois Hudal.

In 1944, Hudal had taken control of the Austrian division of the Papal Commission of Assistance (PCA), which provided assistance to displaced persons. After the war, the PCA would also set up the Nazi "Rat Lines" to assist Nazi war criminals escape justice.

After hiding for about five months, Bormann finally got up enough nerve to visit Munich, the Bavarian capital around October 1945. He was later spotted by someone who had no love for him on July, 1946 and was reported to the Authorities. So, he had to leave Munich and go back into hiding. In the early summer of 1947, Bormann once again was on the move and would soon be in Italy.

Using Bishop Hudal as an intermediary, Evita Peron arraigned a meeting with Martin Bormann in an Italian villa at Rapallo, Italy. There, Evita told Bormann that she had taken $800 million, which represented three quarters of the funds Bormann had placed in German banks in Argentina, and intended to place the $800 million in Swiss numbered accounts for the Peron's personal use.

Previously, Evita had convinced Bormann's agent in Argentina, Ludwig Freude to place these funds in her and Juan Peron's name to prevent later confiscation by the Allies after the war. Now, she was explaining that this $800 million was a protection payment for all the Nazis hiding in Argentina under her and Juan Peron's leadership.

Bormann had no choice but to accept this "hold up" if he, Hitler and many other Nazis were to stay alive in Argentina. But, in his own crafty mind, he was already planning how to get the money back.

And, the man he would place in charge of recovering these funds would be Otto Skorzeny - Hitler's favorite commando.

After this meeting with Evita Peron, Bormann returned to Bavaria. On October 1946, he had been tried in his absence at the Nuremburg War Crime Trials and found guilty of war crimes and sentenced to death. So, he was keeping a low profile. In his Bavarian redoubt, he had a 200-man well-armed army of former SS protecting him.

On August 17, 1947, a guide led Bormann and his small army on a secret rout over the Alps to a hidden base just north of Udine, Italy. In December, Bormann was ready to move again. But, a member of the British Army's Judge Advocate General's office in Italy, Captain Ian Bell, had been tipped off on Bormann's presence in Italy.

Bell ordered a spotter plane to fly over the area where Bormann was reported to be and verified Bormann and his army's presence.

Two days later Bell called in an air strike on the hidden base. An American aircraft flew over the base and dropped a bomb. The Nazi SS scattered and some were captured. Under interrogation, they revealed that they had been guarding Bormann and that he was planning to escape. They also revealed Bormann's escape route.

Captain Bell and two of his sergeants were waiting where Bormann was planning to travel. Soon, they saw a large black car and two trucks with trailers and about 16 men in the group – 6 in each truck and three in the black staff car with Bormann. This was too large a group for Bell to take on. So, Bell and his two sergeants followed the convoy at a safe distance. Finally, Bell came across a phone booth and called in for instructions.

He was shocked by what he was ordered to do. His commanding officer ordered, "Follow, but do not apprehend, now I repeat, do not apprehend".

Bell and his two sergeants followed Bormann's convoy for more than 670 miles. Bormann's convoy passed easily through military road blocks along the way until arriving at the dock of the Italian port of Bari. Bell watched as the trucks and trailers were loaded by crane aboard an awaiting ship and as Bormann walked up the gangway to board the ship. Finally, the ship sailed away.

Later, in an interview Bell explained that he thought that the Vatican had used their influence with the Italian government to allow Bormann's escape.

Bormann arrived at Buenos Aires on May 17, 1948 aboard the *Giovana C* from Genoa – a different ship than the one departing Bari. He was dressed as a Jesuit Priest and traveling with a

Vatican Passport under the name, Reverend Juan Gomez.

Later that same year, Bormann would meet with Hitler at the Inalco estate. Again, Bormann was dressed as a Catholic priest, this time calling himself Father Agustin. This visit lasted a little over a week and then Bormann returned to Buenos Aries to carry on business. As long as the Perons were in power, he felt fairly safe in that center of commerce.

By January 1949, Otto Skorzeny was also in Buenos Aries. His mysterious escape from the Darmstadt Interment camp had been assisted by the CIA, whom Skorzeny was now working for alongside Gehelen. Soon, he would soon meet with the Juan and Evita Peron.

He didn't even bring up the subject of the $800 million Evita had taken from the Aktion Alderflug funds that Bormann had deposited in Argentinian banks - although his ultimate mission was to recover these funds.

He instead discussed how Juan Peron could improve military security in Argentina. He also practically ignored Evita's sexual charms.

Soon, Juan Peron was improving military security around important government facilities and improving interrogation techniques at his intelligence agencies with Nazi efficiency, while Skorzeny became a frequent visitor of the Perons.

Before long, Skorzeny's German spies had also discovered an assassination plot against Evita while she was due to give a talk at one of her many charities. He had his agents arrest the two assassins and hold them at the building where Evita was due to speak.

He warned Evita about the plot. She laughed at the warning, which she didn't believe, and said that she was still going to the speech - but he could come along to protect her. Skorzeny joined her in the chauffeur driven car. When they approached the building, he ordered the car to stop, drew his pistol and ran into the building.

Soon, Skorzeny returned with the two assassins at gun point. They were soon turned over to the police and Evita considered Skorzeny her hero.

A secret love affair soon followed between the two. Skorzeny and Evita would take 3-day tours supposedly to inspect different government installations while actually spending time together at one of Evita's secret residences.

By early 1950, Evita and Juan Peron had given Skorzeny about $200 million of the money they had taken from Bormann. Skorzeny immediately funneled these funds back to Madrid to finance his ODESSA operation with full agreement from Bormann. ODESSA was organized to help former members of the Nazi Party.

Tragically, Evita had cancer and by July 26, 1952, she had passed away. After Evita's death, everything politically went downhill for Juan Peron. She was the film and entertainment celebrity that had boosted Juan into the Presidency to begin with. Her charity foundations had won her husband many votes from the poor class. Also, Juan Peron had made many enemies during his somewhat ruthless rule of the country.

By 1955, the Peron government of Argentina was in a state of collapse and many were seeking to kill Juan Peron. In September 1955, Otto Skorzeny helped Juan Peron escape to a Paraguayan gunboat. Later, Skorzeny helped relocate Juan Peron to safety in exile in Spain.

The price for Skorzeny's services to Juan Peron? The remainder of the Bormann funds.

After the fall of Peron, it was no longer considered safe for Hitler to stay at Inalco. Besides no longer having the protection of a Nazi sympathetic Peron government, there was also the problem of Carlos de San Bariloche becoming a favorite tourist destination. Many of these tourists would also like to take fishing excursions on Lake Nahuel Huapi and sometimes end up fishing in front of Hitler's estate at Inalco.

Bormann gave orders to have Hitler moved from Inalco to a more remote location in the Patagonian countryside called Santa Clara. The fading and frail Hitler became nothing more than a distraction for the international businessman, Martin Bormann. Hitler was placed into a further exile within an exile.

By this time in 1955, Eva and Hitler's Children had been sent to first, to Lisbon, Portugal to be educated in Europe and then later in 1951, to Los Cruces, New Mexico to be educated in the United States. Nazi intelligence officer, Angel Alcazar de Velasco was one person ordered to keep an eye on the children and to occasionally photograph them so Hitler and Eva would know how they were progressing as they grew up. Hitler was now 66, Eva 43, Ursula Hitler, 16 and Adolf Hitler Jr. 10 years of age.

Angel Alcazar de Velasco turns up many times in this story. He gives a different account on how Bormann got to Argentina. Angel Alcazar claims in the book, *Escape from the Bunker*, edited by Harry Cooper, founder of Shark Hunters, that he left with Martin Borman on May 7, 1946 from Villagarcia, Spain on an unnamed German U-Boat commanded by a Captian Jui and spent 18 days underwater on their voyage to Puerto Coig, Argentina.

The account offered in my history comes from the well-researched book *Grey Wolf*. After looking into the matter at length, I feel that Velasco was padding his resume with his version of how Bormann arrived in Argentina. German U-boats could not have made the Voyage from Spain to Argentina underwater in just 18 days. Underwater, typical German U- Boats had a speed of around 5 Knots. At this rate in 18 days the U-boat could have traveled from 2,500 to 3,500 miles – not the over 7,000 miles involved.

Alcazar's account on Hitler's escape from the bunker is also much different than the account offered in the, well researched, *Grey Wolf*. Intelligence officers are known to also deal in disinformation.

After the move to the remote and humbler Santa Clara estate, Hitler became more depressed. According to his doctor, Dr. Otto Lehmann, Hitler also was suffering from Parkinson's Disease, which made his actions tremble. Between the years 1957 to 1961, Hitler's health suffered a serious decline, according to Lehmann.

On, February 13, 1962 at the age of 72, Hitler passed away of natural causes at the Santa Clara estate, in Argentina. Martin Bormann would pass away in 1978. Although the CIA would help Otto Skorzeny fake his death in 1976, Otto Skorzeny would actually live in Florida where he was known as "Big Ed" until 1999. Eva Hitler actually lived into the 21 Century!

The Cold War

The Plan "B" for the third Reich, besides getting the U.S. to continue their war against Russia, resulting in the Cold War, was to relocate their industries to neutral countries and to rebuild and strengthen them after the war and to also quietly take over the U.S. industrial might.

During the Nuremburg war crimes trials, the I.G. Farben chemical cartel, which included BASF, Bayer, Hoechst, and other German chemical and pharmaceutical companies was investigated. Without I.G. Farben, Germany would not have been able conduct World War II. It was *the* major supplier of fuel, explosives, chemicals and technology needed to operate the German war machine.

As documents show, IG Farben manufactured the Zyklon B poison gas used at Auschwitz and was intimately involved with other human experimental atrocities committed by Mengele at Auschwitz.

The cartel was broken up and twenty-four Farben executives were sent to prison for crimes against humanity. However, in a matter of just 7 years each of them was released and began filling high positions in each of the former Farben companies. Overseeing their early release was U.S. High Commissioner for Germany and Knight of Malta, John J. McCloy, who formerly served as U.S. Assistant Secretary of War.

Like Allen and John Foster Dulles, McCloy was a Wall street lawyer. And like the Dulles brothers, he did a lot of legal work for corporations in Nazi Germany. He even advised the major German chemical combine I.G. Farben. So, his early release of the war criminal executives at I.G. Farben was not so surprising.

But, McCloy also released many other Nazi war Criminals after only serving a portion of their sentences. As a Knight of Malta, McCloy was influenced in these decisions by Jean-Baptiste Janssens, at that time the Jesuit Superior General at the Vatican. Interestingly, McCloy was also the brother- in-law of post war Chancellor of West Germany and Knight of Malta, Konrad Adenauer.

Even after I.G. Farben was broken up after the war, the companies comprising this cartel would later prosper. And many of these executives went on to again head these companies as if nothing had ever happened.

Today (2017), Bayer for example, is much larger than the entire I.G. Farben Cartel was during World War II! And the ex-Nazi company, Bayer, has just brought out Monsanto for $66 Billion. This buy out occurred within months of Monsanto merging with Acadmi (formerly Blackwater). That merger provided enhanced security for the increasingly demonstrated against, earth destroying company, Monsanto. And the rumor on the street is that Bayer will use Monsanto's expertise with genetic modification to create a patented GMO form of cannabis for that emerging lucrative market. (10)

While, the Nazi influenced, CIA was promoting the Cold War with Russia, another intelligence agency more loyal to American ideals, was secretly cooperating with Russia to jointly defend the planet from a possible extraterrestrial invasion. That agency was Naval Intelligence. The project was a secret joint space program whose jurisdiction would be limited to space and not to be used against any Earth nation.

National leaders understood the value of having an "enemy" to unify the people with their government that would protect the people from this "enemy". This would justify an expensive arms race on both sides. And money from this arms race would be siphoned off to fund the secret joint space program that we will hear more about later.

In the first place, all parties understood that with the event of nuclear and thermonuclear weapons, an all-out war would benefit no one, and could even destroy all life on Earth.

The idea was to keep the war a cold one, with military contests to be fought in smaller non-nuclear countries. And many, totally unnecessary, wars, having absolutely nothing to do with national defense, were fought merely to keep the defense contractors profitable.

The Vietnam war was a typical example. The Vatican desired to create a Catholic country in South Vietnam after the French defeat at the Battle of Dien Bien Phu by Vietnamese independence fighters in 1954.

A Paris peace agreement divided North Vietnam from South Vietnam, with Catholic, Emperor Bao Dai ruling the South and Ho Chi Minh, the North. Elections would be held in 1956 to elect one leader to rule a unified Vietnam. Bao Dai had appointed, the staunch Catholic, Ngo Din Diem to be Prime Minister of South Vietnam in 1954.

The Catholic missionaries in North Vietnam told their congregations that Ho Chi Minh was a communist that would persecute Catholics and persuaded them to move to South Vietnam. Catholic farmers were told that they would be provided with land in the south. These factors led to a massive exodus of these Catholics to South Vietnam.

In actuality, Ho Chi Minh had modeled the constitution for Vietnam closely after the U.S. Constitution, including a freedom of religion clause. So, the Catholics were basically using propaganda to further their own agenda.

In any case, even after this large Catholic migration, the Catholics represented only about a third of the population in South Vietnam, the rest being primarily Buddhist. Bao Dai was considered a playboy that spent most of his time in Paris and was soon replaced by Ngo Din Diem in a heavily rigged referendum in 1955. Diem started a program of religious persecution of the Buddhists. Land was taken from them and given to the Catholic farmers arriving from the North. Buddhist monasteries were attacked and Buddhist protesting unfairness were imprisoned.

Meanwhile, the Catholic lobby, led by New York Cardinal Spellman, pressured the Eisenhower Administration to not allow the 1956 Vietnamese elections to be held. They desired Vietnam to remain divided and knew that Ho Chi Minh would win the election by an overwhelming margin. This would cause the Catholics to lose their advantage in South Vietnam.

This refusal to hold the promised election caused Ho Chi Minh to organize a military invasion of the South and soon Diem was requesting help from the U.S. Soon, the U.S. was sending South Vietnam military aid and advisors to train the South Vietnamese military. This military was largely comprised of Buddhists who saw little reason to help the Diem government that was persecuting their fellow Buddhists. Often when encountering the enemy these soldiers would simply disappear into the jungle. (11)

When U.S. President Kennedy discovered that Diem was actually using some U.S. military aid to attack Buddhist monasteries, he threatened to withhold U.S. military aid. Eventually Kennedy decided that Diem had to be replaced. When that didn't seem to help South Vietnam's cause, Kennedy decided to withdraw U.S. support for a lost cause.

However, the Military Industrial Complex and other corporations like oil and banking had other plans. They wanted to escalate the military involvement in Vietnam – not withdraw. So, Kennedy was assassinated and replaced by a crony of these corporations - L.B. Johnson. Thus, started the major escalation of U.S. involvement in Vietnam in 1964, which lasted until the defeat of South Vietnam in 1975. After that, Vietnam was finally united as a country independent of other countries interests.

Monsanto was one company which profited handsomely from the Vietnam war. But, in 1979 a lawsuit was filed against Monsanto and other manufacturers of agent orange, a defoliant used during the Vietnam War. Agent orange contained a highly-toxic chemical known as dioxin, and the suit claimed that hundreds of veterans had suffered permanent damage because of the chemical. In 1984 Monsanto and seven other manufacturers agreed to a $180 million settlement just before the trial began. With the announcement of a settlement Monsanto's share price, depressed because of the uncertainty over the outcome of the trial, rose substantially.

Monsanto has had other legal problem with their products. In 1929, Monsanto began production of PCBs (polychlorinated biphenyls) in the United States. PCBs were considered an industrial wonder chemical - an oil that would not burn, was impervious to degradation and had almost limitless applications. Today PCBs are considered one of the gravest chemical threats on the planet. PCBs, widely used as lubricants, hydraulic fluids, cutting oils, waterproof coatings and liquid sealants, are potent carcinogens and have been implicated in reproductive, developmental and immune system disorders. The world's center of PCB manufacturing was Monsanto's plant on the outskirts of East St. Louis, Illinois, which has the highest rate of fetal death and immature births in the state.

Monsanto produced PCBs for over 50 years and they are now virtually omnipresent in the blood and tissues of humans and wildlife around the globe - from the polar bears at the north pole to the penguins in Antarctica. These days PCBs are banned from production and some experts say there should be no acceptable level of PCBs allowed in the environment.

In 1995, Monsanto ranked 5th among U.S. Corporations in the EPA's Toxic Release list. Having discharged 37 million pounds of toxic chemicals into the environment, Monsanto was ordered to pay $41 million to a waste management company in Texas due to concerns over hazardous waste dumping.

Also in 1995, Genetically engineered canola (rapeseed) which is tolerant to Monsanto's Roundup herbicide was first introduced to Canada. Today 80% of the acres sown are genetically modified canola. In 1996, Monsanto introduces its first biotech crop, Roundup Ready soybeans, which tolerate spraying of

Roundup herbicide, and biotech BT cotton engineered to resist insect damage.

As Monsanto had moved into biotechnology, its executives had the opportunity to create a new narrative for Monsanto. They began to portray genetic engineering as a ground-breaking technology that could contribute to feeding a hungry world. Monsanto executive Robb Fraley, who was head of the plant molecular biology research team, is also said to have hyped the potential of GMO crops within the company, as a once-in-a-generation opportunity for Monsanto to dominate a whole new industry, invoking the monopoly success of Microsoft as a powerful analogy.

But, according to Glover, the more down-to-earth pitch to fellow executives was that "genetic engineering offered the best prospect of preserving the *commercial* life of Monsanto's most important product, Roundup in the face of the challenges Monsanto would face once the patent expired."

Monsanto eventually achieved this by introducing into crop plants genes that give resistance to glyphosate (the active ingredient in Roundup). This meant farmers could spray Roundup onto their fields as a weed killer even during the growing season without harming the crop. This allowed Monsanto to "significantly expand the market for Roundup and, more importantly, help Monsanto to negotiate the expiration of its glyphosate patents, on which such a large slice of Monsanto's income depended." With glyphosate-tolerant GMO crops, Monsanto was able to preserve its dominant share of the glyphosate market through a marketing strategy that would couple proprietary "Roundup Ready" seeds with continued sales of Roundup.

The major health problem with this business strategy is that

Roundup contains glyphosate which is a known cancer causing chemical. This Roundup eventually gets into the ground water and into rivers, lakes and the oceans, affecting all life that it comes in contact with.

In addition to this serious problem is the fact that Roundup Ready GMO plants also contain glyphosate in their cell structure. So, no matter how much you wash these plants you are still going to eat glyphosate. This helps to explain the increasing rate of cancer in the U.S. and around the planet.

Ideally, Monsanto should have been outlawed for contributing to the ecological destruction of the planet. But, in a capitalistic nation like the U.S., money talks and consequences be dammed!

Mind Control

In the book 1984, George Orwell warns that people are in danger of losing their human qualities and freedom of mind without being aware of it while it is happening because of psychological engineering. We have learned to expect the Soviet Union and the People's Republic of China to use "mind control" on its citizens, but not the "free world."

But, why shouldn't the so called free world, which is ruled by the financial elite, also have institutions that would use mind control to program the people to support an agenda that this financial elite would desire? As we shall see, there is just such an institution in the United States as well as the U.K. and E.U. which make up most of the free world.

Also, simple mind control techniques are often used in every day advertising on television, radio, internet and the printed media. Somebody is always trying to sell us something, whether it is a product, religion, disinformation or a political ideology.

The most effective way to conquer a man is to capture his mind. Control a man's mind and you control his body. Most people don't pay conscious attention to the things that affect them subconsciously.

The best way to counter mind control is to recognize the techniques being used and how they work. So, let us explore some of the techniques so that we can recognize them when they are being applied to us.

What the conscious mind believes, the subconscious acts on. It works like programming a computer. You feed information into a computer, and the computer acts on it. However, if the information you feed into the computer is wrong, it still acts on it! If you give yourself incorrect information or if others give you incorrect information, the memory banks of your subconscious mind do not correct the error but act on it!

One of the best techniques is to turn off the conscious thinking mind by a distraction to access the subconscious mind. Distraction focuses the attention of the conscious mind on one or more of the five senses in order to program the subconscious mind.

The car advertisements often use a beautiful woman standing beside a car to help sell it. This woman is a distraction to turn off the thinking mind. Now you are having dreams of being with the beautiful woman instead of thinking about things like how reliable is the product, how good is the mileage, how much will it cost etc. To sell products to women, the advertisement might have a handsome man involved. You get the idea.

Another technique is repetition. A current example is the corporate news media, continually hammering us with the repetitive meme of Russian interference in our election with no convincing proof being offered. This is the technique of the "Big Lie" that propaganda minister of Nazi Germany, Goebbels used. Repeat the story enough and the less discerning members of the public will start to believe it. Repetition of the information imbeds it in your subconscious mind so that your acceptance of its truth (accuracy) becomes a conditioned response.

Often, people will be confused when they hear different versions of a story from different sources. If you accept the information as true, it is cataloged that way. And if you reject the information as false, it is cataloged that way. However, if you don't know if the information is true or not, your trust in the source of information determines whether you accept it, even if you are not sure or don't understand it.

Often in this case, the person will accept the story of the source that in their mind seems the more believable. And, a newscaster is an accepted and respected authority figure, thus encouraging acceptance of the information as believable.

Early after its creation in 1947, the CIA recognized this fact. Operation Mockingbird was a secret campaign by the United States Central Intelligence Agency (CIA) to influence media. Begun in the 1950s, it was initially organized by Cord Meyer and Allen W. Dulles, it was later led by Frank Wisner after Dulles became the head of the CIA.

The organization recruited leading American journalists into a network to help present the CIA's views, and funded some student and cultural organizations, and magazines as fronts. As it developed, it also worked to influence foreign media and political campaigns.

In Orwell's 1984, the primary means of oppression is the absolute control of information. And this information plays a large role in the mental conditioning of the masses which creates their "reality".

Much of the "news" we receive is created in CIA offices and then fed to the major news media corporations, like the *New York Times, CNN* and *Washington Post*, which are fully onboard with Project Mockingbird. A good example of this CIA generated news was the "weapons of mass destruction held by Saddam Hussein in Iraq" story used to justify the U.S. 2003 Invasion of Iraq. The "Russia Hacked our election" story is another Project Mockingbird story, to justify further anti-Russian Sanctions.

Actually, the covert operation, Project Mockingbird, being used in the United States is illegal. This is because the CIA by law, is not supposed to conduct covert operations within the United States.

But, the CIA has a large disrespect for the law, as I thoroughly document in my book *CIA: Crime Incorporated of America.* The CIA has been the largest illegal narcotics trafficker in the planet. And, the CIA has been the organizer and supporter of terrorist organizations like, Al Qaeda, ISIS, and the Muslim Brotherhood. The CIA assassinates political leaders like Ngo Dinh Diem of South Vietnam, Trujillo of the Dominican Republic, U.S. President John F. Kennedy and many others. So, for all intents and purposes, the CIA *is* above the law.

Another method of fake news is just to ignore news. I used to wonder why, I could read, well documented, books about CIA illegal narcotics trafficking but never hear a word about this hot news story in the press or T.V. news. Apparently, this is a forbidden subject to a news media that works in close cooperation with the CIA.

This control of news by intelligence agencies also occurs in the U.K. and the E.U. and for the same reasons. The British MI-6 Germany's BND and the CIA collude on many of their operations, including control of the news.

Dr. Udo Ulfkotte, famous German newspaper editor of *Frankfurter Allgemeine Zeitung,* on Oct. 7, 2014, went on public television stating that he was forced to publish the works of intelligence agents under his own name, also adding that noncompliance with these orders would result in him losing his job.

He felt that he was betraying the German people and no longer, in good conscious, could continue. He resigned from his prestigious job and embarked on a public speaking program to educate the public on what was really going on.

"The German and American media tries to bring war to the people in Europe, to bring war to Russia. This is a point of no return, and I am going to stand up and say… it is not right what I have done in the past, to manipulate people, to make propaganda against Russia."

Ulfkotte said most corporate media journalists in the United States and Europe are *"so-called non-official cover,"* meaning that they work for an intelligence agency. *"I think it is especially the case with British journalists, because they have a much closer relationship. It is especially the case with Israeli journalists. Of course, with French journalists. … It is the case for Australians, journalists from New Zealand, from Taiwan, well, there are many countries,"* he said.

Ulfkotte also wrote the book *Brought Journalists,* which became a best seller in Germany. That is when he started getting threats. Acknowledging that his life was under threat, Ulfkotte explained that he was in a better position than most journalists to expose the truth because he didn't have any children who could be threatened.

German media, who were banned from reporting on his work or his book in recent years, reported that he died of "heart failure at the age of 56".

It turns out that mind control and controlling the media has been going on for most of the twentieth century. The Rothschilds owned both the Associated Press and Reuters, the primary source of news to the news media. And that source of news was used to propagandize the United States public into entering World War I.

To turn on the television these days to watch the political, national and world "news" is a surreal experience. You have a dissociative experience as the presenters present not "news" but carefully crafted scripts inventing scenarios out of whole cloth that have nothing to do with what is really happening in any given situation. Even the weather "news" is molded to keep mention of abrupt climate change to a minimum and sports "news" seems like a rehearsal for war news to come as our team smashes their team with our catastrophic weapons, erases them from the earth, never to play again.

But whatever form the propaganda takes, it is a crime against the people, a crime against the republic, a crime against democracy, and since it is a part of the hybrid warfare campaign being conducted and because it is used to provoke a large aggressive general war, it is a war crime.

It is a crime against the people because the people are in essence the state, the nation. The leaders of our nations are merely our representatives placed in positions of power through various, more or less "democratic" mechanisms to act for our benefit, on our behalf. But when these leaders instead represent secret cabals of financiers and industrialists who want to use the government machinery for their private benefit against the interests of their people then they have betrayed the people, have sold them out to the highest bidder. Their lies flow from this betrayal for if their wars were just, they would not need to use propaganda. But their wars are not just, they are the actions of gangsters writ large and so to get the people to go along, to fool them, they, by necessity, have to lie to the people.

It is a crime against democracy for the same reason, for democracy means that representatives of the people put in positions of power have a duty to inform the people honestly on all issues, to present all the facts and arguments, and most importantly, fulfil their duty to preserve the peace and to seek peaceful resolutions of differences between nations. But again, their wars for the profit of a few are always against the interests of the people and so the lies become part of the system of control, and with each lie the grave of democracy is dug deeper and deeper.

It is a crime against the republic because the republic is the people ruling themselves, in the name of the people, not the people ruled by a monarch or emperor, who rule in their own name. So when the leaders of a republic lie to the people of the republic they repudiate the republic and act against its interests and for the private interests of those who control them. They subvert the republic and destroy it.

The ideology of American foundations was created by the Tavistock Institute of Human Relations in London. In 1921, the Duke of Bedford, Marquess of Tavistock, the 11th Duke, gave a building to the Institute to study the effect of shellshock on British soldiers who survived World War I. Its purpose was to establish the "breaking point" of men under stress, under the direction of the British Army Bureau of Psychological Warfare, commanded by Sir John Rawlings-Reese.

The Tavistock Institute is headquartered in London. Its prophet was Sigmond Freud. Tavistock's pioneer work in behavioral science along Freudian lines of "controlling" humans established it as the world center of foundation ideology. Its network now extends from the University of Sussex to the U.S. through the Stanford Research Institute, Esalen, MIT, Hudson Institute, Heritage Foundation, Center of Strategic and International Studies at Georgetown, where State Dept. personal are trained, US Air Force Intelligence, and the Rand and Mitre corporations.

The personnel of the corporations are required to undergo indoctrination at one or more of these Tavistock controlled institutions. A network of secret groups, the Mont Pelerin Society, Trilateral Commission, Ditchley Foundation, and the Club of Rome is conduit for instructions.

Tavistock Institute developed the mass brain-washing techniques which were first used experimentally on American prisoners of war in Korea. Its experiments in crowd control methods have been widely used on the American public, a surreptitious but nevertheless outrageous assault on human freedom by modifying individual behavior through topical psychology.

A German refugee, Kurt Lewin, became director of Tavistock in 1932. He came to the U.S. in 1933 as a "refugee", the first of many infiltrators, and set up the Harvard Psychology Clinic, which originated the propaganda campaign to turn the American public against Germany and involve us in World War II.

In 1938, Roosevelt executed a secret agreement with Churchill which in effect ceded U.S. sovereignty to England, because it agreed to let Special Operations Executive control U.S. policies. To implement this agreement, Roosevelt sent General Donovan to London for indoctrination before setting up OSS (now the CIA) under the aegis of SOE-SIS. The entire OSS program, as well as the CIA has always worked on guidelines set up by the Tavistock Institute.

Another prominent Tavistock operation is the Wharton School of Finance, at the University of Pennsylvania. A single common denominator identifies the common Tavistock strategy---the use of drugs. The infamous MK Ultra program of the CIA, in which unsuspecting CIA officials were given LSD, and their reaction studied like "guinea pigs", resulted in several deaths.

The U.S. Government had to pay millions in damages to the families of the victims, but the culprits were never indicted. The program originated when Sandoz AG, a Swiss drug firm, owned by S.G. Warburg Co. of London, developed Lysergic Acid [LSD]. Roosevelt's advisor, James Paul Warburg, son of Paul Warburg who wrote the Federal Reserve Act, and nephew of Max Warburg who had financed Hitler, set up the Institute for Policy Studies to promote the drug. The result was the LSD "counter-culture" of the 1960s, the "student revolution", which was financed by $25 million from the CIA.

One part of MK Ultra was the Human Ecology Fund; the CIA also paid Dr. Herbert Kelman of Harvard to carry out further experiments on mind control. In the 1950s, the CIA financed extensive LSD experiments in Canada. Dr. D. Ewen Cameron, president of the Canadian Psychological Association, and director of Royal Victorian Hospital, Montreal, received large payments from the CIA to give 53 patients large doses of LSD and record their reactions.

Today the Tavistock Institute operates a $6 Billion a year network of Foundations in the U.S., all of it funded by U.S. taxpayers' money.

Ten major institutions are under its direct control, with 400 subsidiaries, and 3000 other study groups and think tanks which originate many types of programs to increase the control of the World Order over the American people.

The Stanford Research Institute, adjoining the Hoover Institution, is a $150 million a year operation with 3300 employees. It carries on program surveillance for Bechtel, Kaiser, and 400 other companies, and extensive intelligence operations for the CIA. It is the largest institution on the West Coast promoting mind control and the behavioral sciences.

Secret Space Programs

After the war, the U.S., Russia, Canada, and Briton got much of the German technology. The German patent office was raided and literally tons of paperwork with the patents were confiscated and transferred to Wright Patterson Army Air Force Base AMC (Air Material Command) Alien Technology Division, for analysis. This was in addition to all the other sources of intelligence on German Technology.

Members of the German engineers that built the Mittelwerk were sent to Wright Patterson to design underground facilities there to house the recovered German and extraterrestrial technology. These men included Xaver Dorsch who replace Fritz Todt of Germany's Todt Organization which designed many of Germany's underground facilities.

From these underground bunkers, this technology was back engineered, largely by Operation Paperclip Nazi engineers and scientists brought over from Germany. These included men like Siegfried Knemeyer, the former head of the German RLM (The Reichsluftfahrtministerium), the Third Reich's Air Ministry for aircraft development for the Luftwaffe, and Dr. Hans Amtmann, an expert in vertical takeoff aircraft, and Dr. Alexander Lippisch who was well known and a pioneer in tailless aircraft, the US Delta wing fighter, the F-102A Delta Dagger and an advanced design of a ground effect flying boat.

These men also did a reverse engineering on the object that crashed near Roswell, New Mexico as well as recovered German saucers. It was not that our scientists and engineers were incapable. It was because the German technology was based on "other science" developed in Germany, like Victor Shauberger's *Repulsine* levitation science, Karl Schappeller's glowing magnetism and primary state of matter, or torsion field mechanics which could create levitation and inertia shielding around a craft, which was incomprehensible to our scientists trained in conventional science.

When back engineering some of the extraterrestrial craft, the engineers and scientists discovered a particularly amazing thing. The material of the craft seemed alive!

Not only was the craft able to self-repair its damage, like a wounded animal could heal its wounds, although faster, but also the craft could respond to and be controlled by the thoughts of its pilot. This amazing extraterrestrial technology seemed to be based on a combination of nanotechnology and biology. It definitely was based on a manufacturing process that amplified the inherent consciousness imbued into matter by ether. These discoveries quite likely led to the science fiction TV series, *Farscape*.

When the Vrill medium, Maria Orsic was giving Dr. Otto Schumann her channeled information to build the JFM series of saucers, at one point she delivered plans for a mental controller headband so that these saucers could be controlled by the thoughts of the pilot. Later, the *Vrill 7 Giest* saucer could be flown telepathically by Maria Orsic - without the headband! So, even the Germans were able to develop some of this technology which they telepathically received from beings on Alderberan in the 1920s.

Between 1942 and 1947, William Tompkins had access to highly classified projects and was involved in some of the most unprecedented advanced scientific programs on the planet. During this time, the infamous Philadelphia Experiment also occurred. William Tompkins and Admiral Rico Botta, were getting intelligence about this experiment gone wrong. And Tomkins stated that Admiral Botta attended several secret meetings, without Tompkins himself present, concerning this experiment.

After the official end of WWII, Admiral Rico Botta went on to play a key role in starting a U.S. Navy led secret space program filling a number of positions until his final "official" assignment at Naval Air Material Center, Philadelphia, from 1950 until his retirement in 1952.

Meanwhile, Secretary of the Navy, James Forrestal, commissioned the highly above top secret, think tank, project RAND in December, 1945 to study the implications of threatening extraterrestrial agendas.

This was because of U.S. recovered crashed extraterrestrial UFOs in 1941, at Cape Girardeau, Missouri, as documented in the book *MO41 The Bombshell Before Roswell,* by Paul Blake Smith, and in 1942, another two recovered UFOs.

Tompkins revealed that in early 1942, after the Battle of Los Angeles UFO incident, the President of Douglas Aircraft Company, Donald Douglas, along with his chief engineer, Arthur Raymond, and his assistant, Franklin Bolbohm, convened an informal working group that included two Navy admirals and two Army Air Force Generals to investigate the research implications of two retrieved UFO craft.

A leaked Majestic Document dating from February 1942, supports Tompkins claim that two UFOs had been retrieved after the Los Angeles incident, by the Navy and the Army respectively:

> "Regarding the air raid over Los Angeles, it was learned by Army G2 that Rear Admiral Anderson has informed the War Department of a naval recovery of an unidentified airplane off the coast of California with no bearing on conventional explanation.

> Further, it has been revealed that the Army Air Corps has also recovered a similar craft in the San Bernardino Mountains east of Los Angeles which cannot be identified as conventional aircraft. This Headquarters has come to the determination that the mystery airplanes are in fact not earthly and according to secret intelligence sources they are in all probability of interplanetary origin."

Also, the Naval Intelligence data coming in from Germany on Nazi secret flying saucer projects and extraterrestrial treaties contributed to the decision to create the RAND project. Project RAND was set up as a highly secret contract to Douglas Aircraft in a highly classified area of Douglas engineering section at Santa Monica, California. Also at the same time, a new office of the Deputy Chief of Air Staff for Research and Development – to which RAND reported - was set up with Major General Curtis Le May as head. On March 2, 1946, RAND was placed under the direction of Douglas's Assistant Chief Engineer, Frank Colburn - giving birth to the Douglas Think Tank, the Advanced Design Section. Frank Colburn had already been researching all he could about UFOs since the battle of Las Angeles, which he had personally witnessed in 1942.

RAND had two missions: (a) to research the potential design, performance, and uses for manmade satellites; (b) to function as a highly classified scientific research program. The latter mission also considered how to effectively deal with extraterrestrial technology that was thousands of years ahead of ours.

On September 24, 1947, MAJESTIC-12 (or MJ-12) was created upon the recommendations to President Truman, from Sectary of Defense, James Forrestal, Director of the Office of Scientific Research and Development, Dr. Vannevar Bush, and Admiral, Roscoe H. Hillenkoetter. These recommendations were prompted by the Roswell and Corona, NM crashed UFO recoveries, as well as the earlier UFO recoveries, which were being examined at the Alien Technology division at Wright Patterson Field in Dayton Ohio.

MAJESTIC-12 was a top secret extraterrestrial related research and development program. The Deputy Chief of Air Staff for Research and Development and RAND both reported to MAJESTIC-12 after its creation. Later, in the Eisenhower Administration, on the advice of the Rockefeller commission to preserve secrecy, MAJESTIC-12 was placed under the control of the CIA. After that, even U.S. Presidents had little knowledge of what MAJESTIC-12 was up to!

On November 1, 1948, the Project RAND contract was formally transferred from the Douglas Aircraft Company to the RAND Corporation, due to conflicts of interests. Project RAND separated from Douglas Aircraft and became the nonprofit RAND Corporation. Some Scientists and Engineers from Douglas went to Rand Corporation and others stayed at Douglass, working at the highly classified, Advanced Design Section of Douglas.

But, another aspect of this breakup between RAND and Douglas was the fact that there were two secret antigravity projects going on – one by the Navy and one by the Air Force. The Navy one stayed at Douglass - even though Douglass was in the aircraft business. And the Air Force one moved to the new RAND location.

Former Sectary of the Navy, James Forrestal, who became the inaugural Secretary of Defense in September 1947, was locked in a number of bitter policy struggles with Stuart Symington, the first Secretary of the U.S. Air Force.

Symington was firmly opposed to Forrestal's policies, and was a direct factor in the events that led to his sacking as Secretary of Defense on March 28, 1949. One issue was that RAND, which Forestall originally created as Secretary of the Navy, was now favoring the Air Force. It was basically a turf war between the Air Force and Navy and their separate secret antigravity space programs.

And significantly, this was the same period when the Navy and Army Air Force collaboration in Project RAND came to an end, and the RAND Corporation was launched under Air Force dominance in early 1948.

These events led to the evolution of two separate and secret space programs – one run by the Navy and one run by the Air Force.

In 1949, William Tompkins started working at Douglas Aircraft corporation. Because of his high security clearances with the Navy, he was placed into Advanced Design Section of Douglas.

According to Tomkins, this secret Think Tank conceived and planned not only the Apollo Manned Moon Mission 4 years before NASA even existed, but also manned planetary missions within our solar system and manned missions to 12 of our closest stars!

As Engineering Section Chief, he conceived dozens of missions and spaceships designed for exploratory missions to the planets that orbit our nearest stars, a station to be built on Mars, a 2,000-man military base on the Moon, and 600-man Naval station for all the habitable planets and their moons.

Obviously, obsolete rocket technology was not to be used on these missions to the stars. Scientists working at Advanced Design knew about the missing terms in Maxwell's equations on electromagnetic interactions. When these missing terms were corrected in the 1950s, the doorway to the stars were opened.

Also, according to Tompkins, Edward Teller also discovered a problem with a time factor in Einstein's $E = MC^2$ or $E = M$ (9×10^{16} Meters^2/Seconds^2). This, Teller corrected in the 1950s. The corrections in Maxwell's EM theory and $E=MC^2$ provided the technology to far exceed the speed of light.

These statements correlate with statements of past head of Lockheed Martin's Skunk Works, Ben Rich. At an alumni engineering meeting at UCLA, in March 1993, Ben Rich said, while showing a slide of a black disc heading towards space:

> "We now have the technology to take ET home. No, it won't take someone's lifetime to do it. There is an error in the equations. We know what it is. We now have the capability to travel to the stars. First, you have to understand that we will not get to the stars using chemical

propulsion. Second, we have to find out where Einstein went wrong."

According to Bill Tompkins, a Navy space craft could leave the Earth and be within the center of our galaxy in about 45 minutes! At Advanced Designs, Bill Tompkins was designing Naval space craft that were about 2 kilometers long for the Naval space program. This program was designed for the Navy – not the Airforce.

This was because the Navy had centuries of experience with voyages that would take months to complete. The planning, provisioning for manned crews living in close quarters for months at a time and so forth was a Navy thing – not an Air Force one.

As it turns out, contrary to Einstein, there are things that travel much faster than the speed of light. Einstein needs to be updated at our universities.

Perhaps the most interesting thing William Tompkins discovered while working at Advanced Design in Douglas Aircraft, was that his secretary, a knockout gorgeous woman, wasn't human at all - but of the Nordic extraterrestrial race! She would never admit it. But, in additional to her female charms, she displayed great intelligence, leadership, and telepathic ability.

Quite often Bill, as she would call him, would hear her thoughts, planted into his mind. And, many of these thoughts provided the perfect resolution to the many problems in design and logistics for the Apollo Moon program and other programs.

Later on, Tomkins would realize that the Nordics were helping the Navy space program and the Draco Reptilians were helping the Air Force space program. In effect, there was a proxy war between the Air Force and Navy in these secret projects. Also, there seemed to be sabotage attempts on the Navy program which were coming from the Dracos, as outlined here: http://exopolitics.org/rand-corporation-part-of-alien-proxy-war-involving-usaf-navy-secret-space-programs/

Along with designing ships for this this secret space program, Tomkins was asked to design Deep Underground Facilities or Deep Underground Military Bases (DUMB), Deep Under Sea Facilities and Submarine Launched Ballistic Missiles (SLBM) technology in the late 1940s and early 1950s. These efforts led to the Polaris nuclear powered Submarines, capable of launching medium range missiles with nuclear warheads. At the time, many in the Department of the Navy thought that SLBM couldn't be done. But, the Germans had already been designing and building SLBM subs at the end of the war.

Also, Bill Tompkins was asked to design an underground tube tunnel system from the Pentagon to Edwards AFB in California and then on to Vandenberg. This system would propel high speed shuttles at supersonic velocities in a vacuum.

The Douglas Advanced Design insiders already were well aware of the extraterrestrial presence around the Earth. Not only were extraterrestrials flying their scout ships in our atmosphere from motherships further out in space, but intraterrestrials were living inside our planet in large underground caverns and in undersea bases which contained whole cities and civilizations unknown to surface humanity. Top people in Military Intelligence were "in the loop" on these extra and intra terrestrials, but it was decided early on that this information must be kept from the public because it would create a destabilizing influence on society.

For these reasons, the Douglass Advanced Design Think Tank people were very careful to conceal the ET subject from the regular Douglas Corporate managers. Bobby Ray Inman was cleared for ET knowledge and acted as a "go between" Advanced Design and Secretary of Defense, James Vincent Forrestal. (12)

Although Forrestal had been opposed to the National Security Act of 1947, and also opposed any form of unification of the Navy with the Army, he had accepted the job of Secretary of Defense at the newly built Pentagon in September of 1947. Secretly, he was also made an original member of MJ-12 by President Harry Truman around the same time. However, he personally disagreed with all the E.T. secrecy and felt that the public had a right to know what was really going on.

His secret membership in MJ-12, which oversaw the ET project and all the corresponding secrecy surrounding that project conflicted with his personal feelings that the public had a right to know about the ET presence on the Earth. And, this began to cause a lot of stress and inner turmoil in his life.

Also, politics entered the scene, As the presidential election grew closer, Forrestal held some confidential meetings with Governor Thomas Dewey, Truman's opponent. These meetings were designed to ensure that Forrestal would continue at the Defense Department if Dewey defeated Truman, as many expected.

Columnist, Drew Pearson exposed the meetings some time before the election. The exposure strained already difficult relations between Truman and Forrestal, and in March 1949, Truman abruptly requested Forrestal's resignation as Secretary of Defense. His membership in MJ-12 was probably also terminated at the same time.

On the very day of Forrestal's removal from office, he alleg-edly went into a deep depressive state. He was flown to Florida where his wife was vacationing.

Dr. William Menninger diagnosed Forrestal as severely de-pressed, but rather than having Forrestal admitted to his own clinic, one that was familiar with such conditions, Forrestal was flown back to Bethesda, Maryland and committed to the VIP suite of the National Naval Medical Center.

This was an odd destination, considering that Forrestal was no longer a government employee. Even stranger was the fact that access to him was severely limited, something the government denied, but that Forrestal's brother, Henry, confirmed.

The Truman Government was concerned that Forrestal would go public with the MJ-12 operation and the government's secret dealings with E.T.s.

Basically, Forestall was murdered on May 22, 1949 and made it look like a suicide by jumping out a window on the 16th floor of the VIP suite at the Hospital.

That gives you a good idea of what lengths the government would go to, in maintaining secrecy about E.T.s and the secret space program.

Nevertheless, secretly, the Nordics extraterrestrials were help-ing the U.S. secret government take the first steps into space. And, William Tompkins was right in the middle of the planning department for this secret space program!

As William Tomkins states, he was working on the Apollo moon landing project at the Advanced Design department of Douglas

4 years before NASA was even created. Their planning called for a 2,000-man Navy moon base and many more moon missions than actually ended up happening. About 400,000 people were directly working on the Apollo program. The Apollo Command module was testing C3I communications systems that were planned to be used on future Navy space ships.

During the final two years of his 12-year employment at the Douglas Aviation Company, Tompkins' innovative designs for planned Apollo missions had greatly impressed Dr. Kurt H. Debus. In July 1962, Debus had become the first Director of NASA's Launch Operation Center (renamed the Kennedy Space Center after the JFK Assassination), a position he held until his retirement in November 1974.

In 1963, Debus appointed Tompkins to a Working Group for the future Launch Operations Center. The two held many confidential meetings over the future of the Apollo program, and discussed its real mission as part of an ambitious Navy space program called "NOVA" for secretly establishing military garrisons on the Moon, Mars and nearby star systems.

The Apollo Moon landings were only the first stage of an ambitious four stage plan for NOVA. Stage 2 of NOVA was to put 10,000 people on the Moon. Stage 3 was to place bases on Mars and other planetary bodies in the Solar System. Finally, Stage 4 was to place manned Navy bases in 12 adjoining star systems.

All of this was going according to plan right up to 1968. Then, a spanner wrench was thrown into all their plans.

The Manned Apollo 8 orbiting lunar Command Service Module (CSM) spotted extraterrestrial space ships closely crossing their orbital path around the moon. Walter Cunningham radioed to Houston Control center, "There is a Santa Claus and a Mrs. Claus too." Houston replied, "You were instructed not to discuss that."

"Santa Claus" was a NASA code word for UFO. Other frequent code words used were "Barbara" for artifacts or any ruins and "go to KILO" meaning make communications less obvious. Commander Shira said "Look at the size of those alien ships! They are coming in fast at 2 O'clock, straight at us." At the time, he was probably too frightened to stick with NASA protocol.

They were debating changing orbit to avoid a collision with the huge oncoming space ship. Then, Cunningham said "No they will pass in front of us" It was decided not to change orbit to avoid the oncoming space ship which narrowly missed the Apollo 8 CSM.

These ET craft were estimated to be about 1.5 Kilometers long by the people at the command center. The Apollo 8 CSM also returned photographs of extraterrestrial bases on the lunar surface. These photos were quickly classified by the CIA as was all reference to ETs or UFOs in the astronaut's messages.

Even so, it was decided to proceed with the manned Moon landing and see what would happen. In 1966 the lunar Orbiter 2, had passed over the Sea of Tranquility at an altitude of 30 miles. The onboard camera photographed 6 pyramids arranged in geometrical patterns. Three were aligned like the Pyramids at Giza, patterning the three stars in the belt of Orion. It was decided the first lunar landing would be in this area.

When the Apollo LEM was preparing to land on the Moon, the crew sent the message, "Oh my God, you wouldn't believe it.

These babies are huge, sir, enormous. I'm telling you there are other spacecraft out there, lined up on the far side of the crater edge, sir. They are on the Moon watching us."

While this part of the transmission was cut off from the public, some Ham radio operators will attest that that is what they heard on their microwave receivers, while listening to the Lunar landing. There were a number of tremendous sized ET space craft parked on the far side of the crater watching them! The covert Navy NOVA plan came to a crushing end during the Apollo 11 mission when Neil Armstrong and Buzz Aldrin were met by a fleet of menacing extraterrestrial spacecraft.

During the Apollo 11 Moon landing in July 1969, Tompkins says he was in the NASA Launch Operations Center at Cape Canaveral as part of a large TRW contingent. He states that television cameras from the Apollo Lander provided a live feed of what was being witnessed by Armstrong and Aldrin.

William Tompkins was at the TRW monitoring station with a live video feed from the LEM. When Armstrong was photographing Buzz Aldrin's space helmet, Tompkins and others present at TRW could see actual ruins of a former civilization on the lunar surface, reflected in his helmet. Then, they heard an alien voice, "finish a total of six of your Apollo missions; take your photos, pick up some rocks, go home and don't come back."

The extraterrestrial occupants of the large star ships intimidating the Apollo 11 mission did not want the U.S. Navy establishing a beachhead for future military bases on the Moon. By preventing the U.S. Navy in moving forward with its plan to put 10,000 people on the Moon using a number of NOVA rocket launchers throughout the 1970's, the NOVA program effectively came to a crushing end in July 1969.

The no trespassing sign had gone up! And, that put an end to Apollo after 1972. And, that is why NASA never went back to the Moon. About 400,000 people were laid off at Douglass, Boeing, Grumman, north American ITT, Caltech, JPL, and numerous other companies involved with the Apollo Program.

It would take the U.S. Navy more than a decade before it could complete the construction of its first antigravity space vehicles as part of a program called Solar Warden. According to Tompkins and other whistleblowers, the first U.S. Navy space battle groups were deployed in the early 1980's during the Reagan Administration, thereby establishing a U.S. Navy presence in deep space for the first time.

Tompkins says that he approached his work by studying the mission parameters for the requested future space battle groups. He then was able to come up with designs that would allow the Navy to fulfill its future space missions.

Creating the configuration of a Naval Space Battle Group comprising kilometer-long vehicles from the mission parameters he had been given, Tompkins explains:

> "I redefined a standard Naval space battle group complement, stating that it would consist of one 2.5-kilometer spacecraft carrier, with a two-star on board as flag, three to four 1.4-kilometer heavy space cruisers, four to five 1-kilometer space destroyers, two 2-kilometer space landing assault ships for drop missions, two 2-kilometer space logistic support ships, and two 2-kilometer space personal transports."

Eventually, there were eight of these space carrier battle groups that were built for the U.S. Navy in the 1980's and 1990's, according to Tompkins.

Although the designing and planning started in 1954 at the highly secret Advanced Design section at Douglas Aircraft for the Navy, many concept models and testing processes were required to finally get a production model created. After his initial designs of the space carriers were completed in the early 1960s, Tompkins claims that it took nearly a decade for detailed architectural plans to be developed, enabling official construction to begin.

Also problematic, was to locate a secret production center that could construct Naval space craft that were kilometers long. One kilometer is 3,280.84 feet. The largest known Navy ship is the super carrier, USS Nimitz, measuring 1,092, feet long and weighing 101,600 tons which was built at the Newport News shipbuilding yard. So, some of these Naval space ships would be over 6 times longer than the Nimitz!

A very large natural cave was discovered in the Wasatch mountains about 100 miles North East of Salt Lake City, Utah. Part of this cave was converted into a naval space ship fabrication factory large enough to build a 2.5-kilometer sized craft. This enormous underground facility has a door which is the size of some 40 city blocks which opens from the surface into the immense underground facility below.

The cave construction site could have been expanded to build 25-kilometer ships if funding was available according to Tompkins. Consequently, building began in the 1970's and the first operational space carriers were deployed in the late 1970's, under a highly classified space program called Solar Warden.

The work was done cooperatively on the giant space crafts by thousands of employees from Boeing, Lockheed, Northrop, and Rockwell corporations.

This noncompetitive cooperation between these different companies was a feature of the fascist efficiency of these companies after they were taken over by the paperclip Nazis in the 1960s and told that the defense of Earth depended on their cooperative working together.

According to William Tomkins, the door mechanism used to launch the gigantic space craft is so precise and seamless that when it opens and closes, anything on the desert floor, including the plants and animal life which may be on that door, is not disturbed nor harmed in the slightest. With the door closed, any hiker in this very remote location could walk right over it and not even know that it was there.

There were parallel programs also going on where smaller space craft were also being built. Newport News shipping yard was used to modify some nuclear submarines being built there. The nuclear reactors were taken out and replaced with antigravity drives. A lot of other equipment was also replaced. These highly modified subs were launched and while at sea could be secretly launched into space.

Also, much smaller saucer like antigravity craft were being built by Northrup Grumman at the "Ant Hill" underground facility in the Tehachapi Mountains near Lancaster, California.

Furthermore, there also was a highly secret cooperation between the U.S. and USSR all during the Cold War in a joint program to defend the Earth from hostile ET invaders. Many of the weapons programs on each side of the arms race had funds siphoned off to finance these secret space programs.

So, the seeming hostility between Russia and the West was advantageously used to convince the citizens of both sides that the astronomical weapons spending was necessary for their survival, while a good percentage of the money to fund these weapons was, via creative bookkeeping, being used to fund these secret space programs that these citizens had no knowledge of.

The Secret Colonizing of Space

The Germans were first of the Earth based governments to create a base on the Moon starting in 1942, according to researcher Vladimir Terziski. They were allowed to by a treaty with the Dracos, using technology contributed by the Dracos. The Moon, is divided up into different sectors by different extraterrestrial civilizations. Therefore, the German base was located in the Draco sector on the dark side of the Moon.

These sectors are used as diplomatic zones for communications between these different civilizations from different star systems in our galaxy. A rigid enforcement system prevents civilizations which might be at war elsewhere, from any combat on the Moon. Most of these bases are primarily underground for protection against solar radiation.

One reason why the moon was chosen for this purpose is that our sun has a natural traversable worm hole portal system connecting to many of these star systems. This allows rapid and easy travel between our sun and these other stars. A high amount of interstellar traffic occurs in our solar system because of this. Also, the Moon makes an excellent observation platform of the Earth.

These same civilizations prevented the large Navy base on the moon planned by William Tompkins and NASA, which put a stop to the Apollo Program.

However, after the Nazi takeover of the U.S. Military Industrial Complex in 1960, the Nazis started inviting certain members of the U.S. military up to their Lunar base. With the assistance of the Military Industrial Complex, the German lunar base was greatly expanded. Eventually, the Germans relocated to another portion of the Draco sector on the Moon and built another base, which became the center for their "Dark Fleet" which is quite secretive and primarily operates outside our solar system.

These Nazis then handed the built up, original base over to the secret agencies within the U.S. for their own use. These agencies then called this base the "Lunar Operations Command (LOC)." Finally, the U.S. (or secret agencies within the U.S.) had a base on the moon.

While this was going on, the Nazis also merged their secret New Berlin base under the Antarctic ice with these secret agencies within the U.S. After all, now they were all on the same team!

According to William Tompkins, the Navy launched their Solar Warden fleet during the Regan Administration. Reagan had tripled the national debt with his Strategic Defense Initiative (SDI), also appropriately called Star Wars. Much of these SDI funds were funneled into the secret Solar Warden program by using creative bookkeeping.

The idea was that *Solar Warden* provided the Solar System with security to allow trade with friendly extraterrestrial civilizations without the risks of being raided by hostile ET groups. A conglomerate of large corporations on this planet secretly created an interplanetary one, called the Interplanetary Corporate Conglomerate (ICC). Subsidiaries of the ICC were also formed, one being the Mars Colony Corporation. CEOs from known Earth based multinational corporations are selected to become CEOs of the ICC.

This information comes from Corey Goode. He spent many years on a *Solar Warden* ASSR "ISRV" (Auxiliary Specialized Space Research, Interstellar Research Vessel) the *Arnold Sommerfeld*, totally cut off from contact with Earth. To keep boredom away, he would spend much of his free time studying his smart glass pad.

This pad had much of the real history of the Earth, Solar system and the secret space program included. He was allowed access to this information because of his diplomatic duties as an Intuitive Empath. He needed good background information on subjects discussed at these ET diplomatic meetings to do his job.

His access to the information on these smart glass pads and these diplomatic conferences made Corey Goode one of the most informed of the secret space program whistleblowers so far.

Other sources of this information include Bulgarian born, Vladimir Terziski, a student of the Russian Academy of Science, who did considerable research into the World War II Nazi Space program in Eastern Europe after the Collapse of the Soviet Union. He has provided me and others with a lot of information on the Nazi flying saucer projects.

Solar Warden has sub operations called the *Earth Defense Force* and the *Mars Defense Force. The Mars Defense Force* is contracted by the Mars Colony Corporation to defend its assets on Mars.

This leads to the sorry situation that the CEO of the Mars Colony Corporation can give orders to the commander of the *Mars Defense Force.* This led to a disastrous decision to raid a Reptilian Temple after a peace treaty between the Reptilians and Mars Colony Corporation had been agreed to - because a greedy CEO wanted some artifacts at the temple!

The *Mars Defense Force* commander knew it was a terrible idea for a number of reasons, but was outranked. The resulting raid was a disaster. Of the original 976-man four division combat team only 35 survived the raid. Nearly four divisions had been completely wiped out. Randy Cramer was one survivor to tell the story, which is related in *Secret Science and the Secret Space Program* and elsewhere.

Another ugly feature of the Mars Colony Corporation is the use of virtual slave labor. The Nazis used slave labor in their armament factories and underground facilities during World War II and, through the ICC, are still using it on Mars to this day.

The colonists, many of high intellect, were brought to Mars starting in the late 1960s. They were told that a global disaster would soon happen to the Earth and they were needed on Mars to preserve the human race. This caused what later was called the "Brain Drain" in England. But, it was also happening in other countries including the U.S. It was an *Alternative Three* scenario. *Alternative Three* is a science fiction story based on what was secretly going on. (13)

Only persons of the proper IQ and Racial features were selected in this colonization program in typical Nazi fashion. And, everything was done in complete secrecy.

Typically, a bunch of postcards were written up by the prospective colonist before leaving Earth. They would claim on these postcards that they had received a wonderful job in a foreign country. These postcards would be gradually post marked in the different country and sent off to their relatives by agents of the secret recruitment program.

So, the relatives, after reading these postcards, would think the persons were in a different country enjoying their new job, and not worry too much about them. After a few years, they would mostly be forgotten about.

Once on Mars, they were put to work, first building the colonies which were mostly underground, and later worked in the factories to produce trade goods. They were cut off from all news or communication with Earth. They were later told that a cataclysm had actually occurred on Earth that wiped out all life on the planet and that they would never be able to return.

The colonists were paid with wages that provided the basic essentials to live. The wages were paid back out to company stores. So, it wasn't actual slavery since they were getting paid. But, it was the next closest thing. The colonists were basically owned by the Mars Colony Corporation and had no chance of ever leaving.

The Mars Defense Force had their bases of operation separate from the colonies that they were defending. No contact was allowed between soldiers and the colonists. This was because the soldiers came from Earth and knew that no cataclysm had actually occurred. This was according to both Randy Cramer of the *Mars Defense Force* and Corey Goode of *Solar Warden*. Goode also states that he knew of five different colonies on Mars. Goode also explained further:

"The ICC has an entire industrial infrastructure that includes bases, stations, outposts, mining operations and facilities on Mars, various moons and spread throughout the main Asteroid Belt (where a "Super Earth Planet" once existed). They have facilities to take raw materials and turn them into usable materials to produce both complex metals and composite materials that our material sciences have not dreamt of yet. They have separate groups of facilities that produce various types of technologies as well as each facility or plant that produces a specific component of a technology so that those working in the facilities and living in the support colonies/bases do not know exactly what they are producing. Much of the time the components are multiuse and are used in cross over projects. There are facilities on Earth that operate in much the same manner that contribute to the SSP on several levels.

There are other bases on Mars that are controlled by Military/Security groups as well as some scientific outposts. These can be owned and maintained by other SSP Programs but are usually going to report to the ICC on some level since the ICC controls much of the Air Space and Security Operations on and around Mars. Most of the security personnel that are assigned to Mars are assigned to and serve under the ICC. The military groups that will be returning to their previous organizations (SSP Groups) are kept isolated from the population and personnel who live and work on the Colonies, Bases and Industrial Facilities that they protect. They are normally in the rather Spartan outposts that I have described previously in other writings. I had seen a few of these outposts built from the "Ground Up". They were always quite a distance from the main underground colonies, bases and industrial facilities and spread out in a Multi-Teared Perimeter Defensive type of system. There are "Non-Humans" also having bases on the planet. Some of them have been there for some time and have the highly coveted larger lava tube systems that have been built out into base systems that are unimaginably huge and can securely reside millions of inhabitants."

The secrecy around these operations is nothing short of astounding. Men who sign up for these defense forces are required to sign up for a 20-year tour of duty. When this time is served the men or women are returned to the Lunar Operation Command. There, they then have their memories selectively erased of their entire 20 years of duty. Then, they are age regressed by 20 years which requires about 2 weeks with advanced medical technology. Then, they are sent back 20 years in time to the time starting the tour of duty. This policy was called "20 and back". Theoretically they would never know that they had done this tour of duty and their lives could normally continue without revealing any secrets.

When Corey Goode was asked what happens when you got back. Did you remember things? Are a lot of people out there right now who served in these programs but can't remember anything? Corey Good replied:

"Well, yeah. There are many, many, many thousands of people. And many of them are programmed to be triggered by this topic. And some of the biggest skeptics are people that actually participated in this. And, yes, for the most part, the blank slating is very effective. And 3% to 5% of the people, the mind wipes or the blank slating does not work. And it's usually the people that are intuitive empaths that it doesn't work in, because they are not just relying on their memories on their hard drive in their brain. They also have a strong connection with their higher self - sort of like a virtual hard drive - with their light body and higher self. Like, you know the same way people have memories of a prior life? Well, in the birthing process in this chemical brain of ours, you can't possibly remember a life 1,000 years ago. But in your soul memory, you can. So, in these programs, they can erase these chemical memories and electromagnetic memories in your mind and play with them. But they can't affect the memories in your soul body or in your higher self. So, in the 3% to 5% of people that have that connection, they have a very difficult time of affecting those people with screen memories and blank slating"

According to Andrew Basiago, during the 1980s, he had gone to Mars a number of times from a teleportation chamber called a Jump Room in El Segundo near Los Angeles Airport. A Number of others in the same program have also come forward and said that they had also gone to Mars via the El Segundo Jump Room.

Michael Relfe, a member of the *Mars Defense Force* also speaks of this Jump Room. VIPs would come to Mars via these Jump Rooms and visit for a period of time, usually less than 2 weeks, and then return to Earth the same way This is according to the *Mars Records* that his wife composed based on Michael Relfe's recovered memories. Relfe and other grunts in the *Mars Defense Force* on the other hand would have to be subjected to the "20 and back." policy. (11)

Another person, Tony Rodrigues, has recently come forward who was recruited into the "20 and back" program.

Tony Rodrigues claims that in 1981, when only nine years old and in 4th grade, he was involuntarily recruited into a "twenty and back" program as punishment for something he did to one of his class room peers.

He publicly embarrassed the son of a high-level member of the Illuminati and Rodrigues says that he was soon after abducted by five aliens. He was then genetically tested to determine what skills he possessed, which could be used in covert 'support' programs and eventually for the secret space programs themselves when he got older.

Rodrigues says that he was first forced to work as a psychic for a drug running operation out of Peru for four years. He would be put asleep with drugs and would psychically determine if a particular drug run would be safe or not. Edgar Casey also was a very accurate psychic that worked while unconscious.

Later he was recruited as an underage sex worker in the area of Seattle from age 13. His wealthy Illuminati owner was a practicing Satanist who hosted group sex orgies and even ritually sacrificed a baby. Rodrigues was actually forced to eat parts of the dead baby.

When he turned 16 years old, in 1988, he was taken to the Moon to be tested for any skills he possessed that would be of most benefit for service in the secret space programs. He was taken to both the Lunar Operations Command or LOC and the trapezoidal shaped Nazi Dark Fleet base, which he observed before landing.

While briefly serving on Mars, he was attacked by a large insectoid and had both his arm and foot bitten completely off. He applied tourniquets and was carried by a Mars Defense Force fighter to the base hospital and passed out. The next morning, he woke up and discovered that he had a new arm and foot and was completely healed!

That hospital technology seemed to be similar to what was described by Randy Cramer, where MDF soldiers blown up in battle could be completely repaired and rejuvenated overnight and ready for battle the next day!

After this incident the project, which Tony Rodrigues was involved with – basically acting as bait to draw the insectoids in so that the Marines special services forces could kill them – was abandoned. At the outpost, the Marines s.s Mars Defense soldiers would always be playing pranks on and teasing members of his group.

Later, Rodrigues was taken to *Aries Prime* and tested to see what abilities he possessed. Then, he was shipped out to Ceres, a planetoid in the asteroid belt, where he worked for 13 years with a German led freighter crew that was part of the "Dark Fleet," initially set up by Nazi Germany during World War II.

When he got off the flight to Ceres and entered the hangar with others who had accompanied him, their group was informed that they were required to obey every order given them. They were told that a few others before them that didn't agree to obey, were shot on the spot. Then, they were asked if any of them needed to be shot. Needless to say, none of that group thought so.

The operation there was very much a Nazi operation. There were swastikas everywhere. Everyone had to give the stiff-arm salute and say Hiel - followed by the name of the leader of the Ceres operation, which would change every so many years. German and English was the common language of the base. But, there were also Extraterrestrials on the base. Also, there was a percentage of women present. Many of the permanent members of Ceres were born there and would probably die there.

Many of the laborers there were mining the mineral resources of Ceres. When some areas were mined out, cities would be built inside the resulting caverns. Others were involved with manufacturing. And trade was carried out with other civilizations both in our solar system and other star systems.

Rodrigues was assigned to ship repair, at first mainly in plumbing, replacing old failed valves. The older ship interior looked like a submarine, with valves and gauges everywhere. Valves and gauges were not threaded but welded into place. Fortunately, he did have automated cutting and welding tools, which made the job easier and produced welds which could pass inspection.

Eventually the old ship was replaced with a newer model and Rodrigues was promoted to the cargo section. Even though his life became improved, he like others of his rank, still had to always wear a collar with which he could be electro shocked if his Nazi superiors thought he needed to be disciplined. Now, he actually would fly with the ship to different destinations and deliver the cargo.

Later, he would visit a base brothel. The women who worked there were young and pretty and were all mind controlled. Before being brought there, their memories were totally wiped out or Blank Slated. So, having no prior memories of their childhood or families, they believed everything they were told. In addition to the men on the Ceres base, many aliens would also visit the brothel and have sex with these unfortunate women.

The Nazi women and men who lived there permanently, always had superior attitude because of their advanced technology and tended to bully people like Rodrigues. These people had to just accept their fate and wait out their 20-year tour of duty, because escape was impossible.

This is merely a short synopsis of the Tony Rodrigues story. He is planning on writing a much more detailed story in book form. (14)

So here we see the gross unfairness in the way corporations are operated. Here on Earth, the CEO's of these corporations receive thousands of times the salaries of the employees. Are they actually worth that much more? It is highly doubtful.

These CEO's have been streamlining their company by firing much of the corporate staff and employees while making the remaining staff and employees work twice as hard. These tactics increase the corporate profit margin which keep the stock prices elevated. But, the company shareholders usually don't see any increase in dividends from these actions because the CEOs give themselves even more pay and benefits like stock options and golden parachutes!

As we can see, the Mars Colony Corporation have a captive work force and the corporate unfairness is even much greater on Mars, where no government regulations are in effect to protect worker's rights. The workers on Mars suffer from a similar plight that the Negros in the Sothern plantations before the Civil War suffered, but in a more modern context.

The trend on this planet unfortunately is heading in the same direction. In the last part of the 20th and early part of the 21st century, we have seen a number of international trade treaties being formed, Like the WTO and NAFTA which removed many jobs from the U.S. where people received a decent wage and moved these jobs to countries where basically slave wages were paid.

There needs to be a means to get our government to bring these Interplanetary Corporations under control and return human rights and decency to their abused employees. The first step is to reveal these secret operations to the public so that they will demand a change. This is a powerful motivation for me to do the research and write this book.

Secret Exopolitics

Exopolitics is the political interaction between Earth based governments and governments based on other planetary systems. These exopolitical interactions have been going on throughout the twentieth century and have been kept quite secret from the ordinary people of the Earth.

The Draco Reptilians, although living inside the earth, have alliances with the extraterrestrial Draco Reptilians on other star systems and also have a base on the Moon. So, they could be considered, in effect, extraterrestrial.

And in the 1930s, these Draco Reptilians entered into a treaty agreement with Hitler's Third Reich and Japan to join in an alliance with them to conquer the Earth. In return, The Third Reich would be given advanced technology, an abandoned underground city in Antarctica and the right to create a base on the Moon. The U.S. didn't have permission to create such a base on the Moon and that is the real reason the Apollo program came to a screeching halt in 1972.

Extraterrestrials from the Pleiades secretly met with President Franklin D. Roosevelt in early 1934. These Pleiadians offered social and economic programs that would transform Earth into a paradise. The only condition was that Roosevelt would have to embark on a disarmament program and give up war. Roosevelt refused and the Pleiadians withdrew from the offer.

President Roosevelt also met with Extraterrestrials known as the Greys. These Greys were from the Orion Star system and were in alliance with the Draco Reptilians. So, they did not have humanities best wishes in mind. The meeting occurred aboard a U.S. Navy ship in the port of Balboa, Panama on July 11, 1934.

These Greys stated that their race once looked like humans. But, atomic wars and pollution killed all the surface life on their planet and they were forced to survive in underground cities. Over millions of years they evolved into their present form with spindly arms and legs and large head with large almond shaped eyes.

Also, they were losing their ability to reproduce. For this reason, they needed to do biological experiments which required human genetics to strengthen their own genetics to regain the ability to reproduce.

For this reason, they wanted permission to abduct human beings for a short time to remove small amounts of genetic material, which would cause no lasting harm to the abducted ones. They possessed the ability to erase the memory of the abducted human, which they promised to do to prevent any emotional trauma of the abduction. And, thy promised to return the abducted ones after obtaining the genetic material.

In return, the Greys promised to give the U.S. advanced technology. Unlike the Pleiadians, they did not require the U.S. to disarm and give up war.

Roosevelt agreed to their terms with the added stipulation that the U.S. government would be supplied with a list of all the abductees. A secret treaty was agreed upon. However, these secret treaties are not lawfully binding because treaties in the United States must be ratified by the U.S. Senate.

Later, when Dwight D. Eisenhower became U.S. president in 1953, Truman handed him the MJ-12 documents and Eisenhower was informed of Roosevelt's treaty with the Greys.

It was decided that the Greys had reneged on parts of the agreement. There were more people being abducted than were on the list to the government. Also, the Greys were holding back on some of the promised technology. This upset the military leaders, who were now aware of the advanced technology the Germans had received from the Reptilians.

By Feburary,1954, an agreement was reached for a meeting between a different extraterrestrial group, called the Nordics by the military, from the Pleiadian star system. This meeting between the Nordics and Eisenhower was to be held at Edwards AFB (then called Muroc Air Field) in California. On February 20, 1954, the meeting took place.

Besides Eisenhower, a number of important persons were present. These included military generals, a reporter for the Hearst newspaper group, Franklin Allen, a Catholic Bishop from Los Angeles, James McIntyre, Gerald Light of Borderland Research, Edward Norse of Brookings Institute and others. All these men had to undergo six grueling hours of interrogation about their private lives and beliefs before being allowed into the meeting area.

The Nordics brought two cigar shaped ships and three smaller flying saucers to the meeting. An E.T. translating device was used at the meeting. The smaller saucers were so light that they could be lifted by hand by the men present, who were allowed by the Nordics to inspect these saucers. The Nordics at one point demonstrated their ability to disappear from plain sight, right in front of the men present. This was an unnerving experience for these astounded humans.

At first, Bishop McIntyre dominated the conversation and asked many questions about the Nordic's religion. They responded that every universe has a God. They stated that they possessed the true record of Jesus Christ and that the first four gospels of the New Testament were in error. They said that Jesus survived the crucifixion and lived to an old age. They explained that hell didn't really exist and the resurrection was just invented by priests to give them more power.

The military generals became impatient with all this religious talk and wished for the Nordics to explain how their technology worked. The Pleiadians refused, stating that humans were not spiritually evolved sufficiently to wisely use the technology they already possessed. They also warned that humankind was presently on a course that would destroy their planet.

They also warned to stop all nuclear testing because it was causing havoc on other dimensions and destroying the planetary ecosystems.

They also stated that they wanted their presence known by all the Earth's people as soon as possible. Eisenhower responded that the public was not yet ready to handle this stupendous knowledge and required time to be prepared for this kind of knowledge.

After this meeting, it was decided to keep everything secret. So, Franklin Allen of the Hearst newspaper group did not write about the meeting he was invited to. However, Gerald Light sent a letter back to Borderland Research, where the above details were derived. Also, Bishop McIntyre later met with Pope Pious XII and transcripts and films of the meeting were stored in the Vatican library. (15)

No actual agreements came from this meeting since Eisenhower was unwilling to stop nuclear testing or inform the people of the Pleiadian presence.

Later, in April 1954, Eisenhower would meet with a different extraterrestrial group from the Betelgeuse star system, which the military would call the Large Nosed Greys. These Greys basically wanted the same arrangements they had with Roosevelt for the same reasons. They would trade advanced technology in exchange for the right to abduct and return selected humans for genetic experimentation. They made no demands on our government to turn away from war or give up nuclear weapons.

The new agreements would include the right to establish underground bases in the U.S. for the Greys to continue their experimentation. Some of these bases would be jointly operated by the Greys and the U.S. military. This latter clause would help insure that the U.S. would get the technology they wanted, which they felt was not being delivered per the former agreement with Roosevelt.

This secret treaty would be called the Greada treaty. This secret treaty was also not actually legal since it was not ratified by the Senate. But, it seemed to give Eisenhower what he wanted – better weapons technology to take on the Germans in their secret Antarctic base.

What he failed to realize was that these Greys were in an alliance with the Draco Reptilians that were allied with these same Germans! So, Eisenhower's plan to obtain weapon superiority over the Germans was, from the outset, doomed to fail.

The famous Roswell UFO crash of 1947 was actually three UFO crashes which marked the beginning of another large and secret exopolitical operation by the U.S. government, later called Operation Crystal Knight by President John F. Kennedy.

Each of the three flying saucers were from the Zeta Reticulum star system on a planet they call Serpo, about 40 light Years from Earth. Each saucer had four crew members aboard.

The cause of these crashes was a new high-powered radar system installed at the nearby White Sands rocket testing center to track the flight of their test rockets. It seemed that these radars interfered with the navigation systems of the saucers from Serpo and caused them to crash in different areas around Corona, New Mexico.

Of the 12 ETs, 10 died immediately in the crashes, one was seriously injured and died within a day and one was in fairly good condition. This one was taken into custody by the UFO recovery teams and given the name EBE1, which was an acronym for Extraterrestrial Biological Entity # 1. All 11 of the ET bodies and EBE1 were taken to the Los Alamos National Laboratories to be studied.

Eventually a form of primitive communication was developed between EBE1 and his hosts at Los Alamos. This communication used pictograms. EBE1 was quite cooperative and serene during the whole ordeal. His vocal cords only allowed high tone sounds and caused the problem with verbal communication.

Later, an operation on his vocal cords allowed a form of speech to develop by EBE1, who eventually was able to communicate in English. He then became quite useful to the scientists trying to discover how recovered equipment from the crashed saucers worked.

One piece of equipment was a communication device which allowed communication with their home planet of Serpo. EBE1 explained how to operate this device and was able to send 6 different communications to Serpo before his death in the summer of 1952. The time lag between sending and receiving these messages amounted to months, which was remarkable enough, since Serpo was 40 light years away. So, replies from EBE1's messages were not actually received at Los Alamos until after EBE1's death.

Eventually, after nine years of trial and error, a system of communication between Earth and Serpo was finally established. Thus, began Operation Crystal Knight during the John F. Kennedy Administration.

In 1962, it was arranged for the Ebens (as we called them) to make another trip to Earth and meet with representatives of our government. The Eben landing was scheduled to take place at Holloman AFB in New Mexico in April 1964.

Kennedy then decided on a cultural exchange program between Earth and Serpo code named Operation Crystal Knight. Part of this program would be to send 12 specially trained human astronaut/diplomats to Serpo for a 10-year period to learn from their culture.

The men chosen for this secret team would be single, with no family – preferably orphans. This, in case they never returned to

Earth. They were severely trained for this operation and their actual identities were scrubbed from the records while new identities were created for them consisting only of 3-digit numbers.

The next Eben landing took place on April 24, 1964 at Holloman AFB as planned. Kennedy was no longer alive by then and now Lyndon B. Johnson was in charge. The Ebens brought a translation communication device with them to facilitate communications. At this meeting, the 12 preserved bodies from the 1947 Roswell crash were given to the Ebens. In return, the true history of the Galaxy, in what was called the Yellow Book, was given to the U.S. representatives there. This Yellow Book could display a holographical display of actual scenes form that history.

The 12-man team, picked to go to Serpo, was waiting in a nearby bus. But, the Ebens wanted to wait another year to prepare before carrying out that part of the project. So, in July 1965 another prearranged landing by the Ebens took place at the Nevada Atomic Test Range 95 miles north of Las Vegas close to Area 51.

This time, the 12-man Serpo team boarded a huge Serpo mothership along with 45 tons of supplies for their planned 10 year stay on Serpo. The Ebens possessed antigravity technology that easily lifted the whole 45 tons of supply into their ship in one operation.

The voyage to Serpo from Earth would take 9 months. Of that 12-man team, only 7 would return 13 years later in 1978. One died on the trip to Serpo of natural causes. Two would later die of natural causes on Serpo and two would decide to stay on Serpo. (16, 17)

For those wondering about the veracity of the above-mentioned sources. It seems that a fellow that claims to have been involved with the original project and worked at the Defense Intelligence Agency (DIA), using the pen name Anonymous, was the primary source.

Even more intriguing, Anonymous also claimed to be an editor of the famous "Red Book", which contains a summary of all the important extraterrestrial contacts between the ETs and the U.S. government from 1947 to the present.

This Red Book is actually brownish orange in color and is well known among highly placed government officials involved in secret UFO investigations and ET contacts. This book is updated every 5 years, as reports come in and are vetted for accuracy and added to a list for review for possible inclusion in the Red Book

Anonymous claims to have served as the editor of the Red Book for a number of editions and to have prepared briefing documents for several U.S. Presidents. He also claims that he isn't working alone, but is a part of a team of 6 DIA employees.

Although he doesn't mention it, based on the level of information he had access to, Anonymous possibly could also be a member of MJ-12. And it seems quite likely that his release of information was authorized by his commander, as is the case with informants like Randy Cramer, who spent 17 years on Mars and 3 years in the Solar Warden space ship *Nautilus* and William Tomkins who helped design these Solar Warden space ships.

At a high level of government, there now seems to be a program of release of former secret information in order to acclimatize the public to the reality of extraterrestrial presence on our planet and official exopolitical interaction with these ETs.

On facet of this acclimatizing process are the many science fiction films based on what is secretly going on, produced by directors like Spielberg who are being advised by military experts with high level security clearances that know about these black projects. And, the other facet are these releases of nonfiction information by people who actually worked on these projects who were ordered to by their superiors to release this information, as with Randy Cramer or given permission to release the information, as with William Tompkins.

Secret Exotic Technology

Those who read this far might think that they are reading a science fiction story. Many of the technological feats mentioned in this history seem to not really be possible and the technology behind these feats seems not to really exist.

Where is the science behind this technology? Why is it not taught in our universities? Why are we still using rockets to get into space if antigravity technology really exists? If free energy is really possible, why are we still using gas guzzling cars?

I have answered these questions in detail in my previous book, *Secret Science and the Secret Space Program*. I will attempt a brief overview of that treatise here.

By the time, I attended the University of California at Santa Barbara in 1979, I was a thirty-nine years old Electrical Engineering Major and had done considerable study of science on my own. In particular, I had studied the experimental work of Nikola Tesla and Thomas Townsend Brown.

Nikola Tesla was the greatest scientist of our time – yet his work was barely mentioned at the Universities. Thomas T. Brown had experimentally discovered the connection between electromagnetism and gravity in the 1920s. He even has patents on his inventions which he called gravitators. His revolutionary work was not mentioned at all at the University where they were still teaching that Einstein was unable to develop a successful unified field theory that unified electromagnetism and gravity.

So, the real question should be "Why isn't the work of these great scientists and inventors being taught at these universities?" I obviously realized there were vast holes in what we were being taught there. It seemed to me that the program I was in was designed to quickly turn out EE graduates for industry to hire. I also minored in physics to get a broader understanding of how the universe works.

I graduated in 1982 because of the extra classes I was taking. In 1985, I started reading books by Thomas Bearden and John Bedini and became exposed to the theory and possibility that free energy was possible.

The basic premise is there are no truly closed systems and the second law of thermodynamics that conserves energy only works in closed systems. For example, we are continually bombarded with neutrinos which penetrate everything. If the energy of these neutrinos could be tapped, we would have plenty of free energy wherever we would be in the universe.

In fact, Dr. Konstantin Meyl claims to tap the energy of this neutrino flux using pancake coils. His web site is here: http://www.meyl.eu/go

The vacuum of space also contains an abundance of zero-point energy. Permanent magnets can access this zero-point energy. Thomas Bearden patented what he called the Motionless Energy

Generator (MEG) – U.S. patent # 6,362,718.

The MEG switches the magnetic flux from a permanent magnet through two different magnetic paths wound with energy pick-up coils. A small amount of input energy is required to operate the flux switch and a much greater amount of energy is picked up by the pickup coils.

In 2005, I created a MEG in my own lab. I experimentally measured a COP of 3, or 3 times more energy out than went in, when the proper resonant input energy was used.

In 1993, I went to a Tesla Symposium at Colorado Springs, Colorado. There I watched Joseph Newman present his energy machine.

Many engineers were present. All were invited to make measurements and check for hidden wires. That is when I knew for a fact that free energy was entirely possible.

Joseph Newman spent 10 years suing the U.S. Patent Office. He would demonstrate his working models and have engineers certify that they worked as claimed. The Patent Office stated that they didn't patent perpetual motion machines. After 10 years of expensive litigation Newman's patent was finally granted.

Since that time, I have observed several other free energy machines in operation and have done extensive research on a number of others.

In Nikola Tesla's later experiments, he worked with high voltage pulsating direct current rather than alternating current. In these experiments, he discovered a whole new realm of electro-science which included the ability to magnify energy rather than merely transforming voltage and current. None of this science is taught at our Universities but can be discovered in the works of Gerry Vassilatos. (18, 19)

After years of my own research, I finally concluded that the science of free energy and antigravity were forbidden subjects among government entities and Universities. There was a very active suppression program going on.

As an example, I will point out the case of Dennis Lee.

In theory, at standard temperature and pressure, each air molecule has an average velocity of 700 miles per hour. That represents a large amount of kinetic energy. Since these air molecules are moving in random directions there is no 700 miles per hour wind. But, this energy presents itself as temperature and pressure. And, heat pumps can extract this energy!

Dennis Lee was a businessman who owned a heat pump factory in the State of Washington. To qualify for an energy tax credit offered by the Carter Administration, Dennis Lee asked his engineers to design a more efficient heat pump. Average heat pumps were then getting a coefficient of performance (COP) of about 4. That meant that these pumps could pump 4 times as much heat energy as the electrical energy the compressors took to run the heat pump.

After a period of experimentation Lee's engineers would develop a heat pump with a COP of 12! This wasn't measured in Lee's company but in independent laboratories. These more efficient heat pumps were manufactured by Lee's company, which qualified for Carter's tax break.

Because of this high COP, Dennis Lee asked his engineers if it would be possible to put the heat into a heat engine driving a generator and get more electricity out than put in. Most of his engineers pointed out the Carnot efficiency of a heat engine limited the efficiency to less than 30% and with friction and other losses it would probably not be possible.

However, one engineer said that he was acquainted with an Australian inventor named Fischer that had invented new type of engine, called the Fischer cycle engine which could have a theoretical efficiency of 90%! This engineer also said that he knew how to contact him. Dennis Lee arranged to meet with Fischer and soon they were collaborating on a new invention.

The secret of the Fischer cycle heat engine efficiency was the use of phase change during the engine's cycle. This engine had a longer piston stroke than a diesel engine. At top dead center, high pressure and high temperature refrigerant would be injected into the cylinder. This would push the piston down by hydraulic pressure. As the pressure reduced, the high temperature fluid would flash into a hot gas, pushing the piston further down while expanding and cooling the gas further. Near the end of the stroke, the cooled gas would condense back to a low temperature and pressure fluid and would exit via special holes in the cylinder.

So, a hot high-pressure fluid would enter the engine and a cooler low-pressure fluid would exit the engine. The difference between the temperature and pressure entering and leaving this engine would be converted into mechanical energy very efficiently.

Meanwhile the electric companies in Washington liked regular heat pumps. But they didn't like the heat pumps manufactured by Dennis Lee's company because they cut the customer's electric bill too much. They got together with the Washington State Attorney General, who brought a law suit against Dennis Lee's company which was forced to shut down because of the law suit.

Dennis Lee felt that things would be better in California and moved his operations to Ventura, California. There, some of his heat pump technology was used to heat the water at the Ventura Junior College swimming pool. But the main thrust of his Ventura operation was research and development.

He organized a presentation of his new energy generator, which used a Fischer cycle engine and his high COP heat pump connected together with an electrical generator. This demonstration was given in an Oxnard auditorium to an audience of about 200 people which included engineers and news reporters.

To start this machine, he had to use an extension cord from the auditorium, but stated that he could have used a battery and inverter combination instead. Once the machine started, he unplugged the machine from the auditorium power and the machine continued running. He invited members of the audience to come up on stage and look for hidden wires. There were none discovered.

Then, he threw another switch which connected the generator output to a bank of lightbulbs which represented a 10-kilowatt load. The bank of 100, 100-watt bulbs brightly lit up as the machine continued to run.

Dennis Lee explained to the audience.:

"This is not free energy, This, is energy coming from the heat energy in the air in this auditorium, which is extracted by our heat pump. As heat is removed, the air cools down and convection currents bring warmer air to our absorber. If this generator runs long enough the air in the whole auditorium will cool down. So, not only are we extracting free electrical energy from the air – we also have an air conditioner."

Within 2 days of this amazing presentation of technology developed by his company. The office and laboratory of Dennis Lee's company was visited by the Ventura County Sheriff Department. The Sherriff Department brought a truck and semi-trailer and started loading it up with machinery from the laboratory and paperwork from the office, including business contracts, and proprietary research documents worth millions of dollars.

Dennis Lee was charged with failing to file an obscure document with the County Clerk's office. Normally, in a case like this, the County Clerk would merely send a letter to the company office. But, the Sheriff's office was desperate to justify their action with this absurd charge. Dennis Lee was held in jail for one year without a trial.

Then, after his company went bankrupt, he was released with all charges dropped. Later, Dennis Lee sued the Ventura County Sheriff Department to return the property they confiscated from his company. They refused, stating that they were keeping it for evidence.

Obviously, the electrical utilities companies did not like Dennis Lee's machine and had enough influence with the authorities to get them to sabotage the efforts of his company in a highly unconstitutional and unlawful act of tyranny.

With the present global climate warming problem, a solution

would be to use massive electrical generators, using the concepts developed by Dennis Lee's company in Ventura California, that would cool the atmosphere while producing electricity with no fuel costs. This is a much better concept than using heat from burning fuel or nuclear reactors to create electricity and much less polluting. Furthermore, there would be no energy storage problems as with wind and solar electrical generation as atmospheric heat is available 24/7.

Another secret exotic technology is technology that can modify the environment or ENMOD technology. Much of this technology, which can affect the weather, steer hurricanes, cause droughts and trigger earthquakes, centers around the High Altitude Auroral Research Project (HAARP) and associated technology.

The HAARP antenna array can direct microwave energy in any desired directions of the sky by changing the phasing between the antennas, which are fixed. This steerable beam of microwave energy can heat up any area of the sky where ions are present, which is normally in the ionosphere.

However, HAARP can also be used in conjunction with aerial spraying - as with chemtrails – which place particulate matter into the air that can interact with microwave energy and cause localized heating below the ionosphere.

This localized heating of the air can affect the weather by changing the jet stream, creating low pressures in one area and high pressure in another, preventing rainfall etc. Other methods of proven weather control use what Wilhlem Riech called Orgone Energy in his Cloud Busters. Some weather engineering companies have overcome droughts in the Mid-East and other areas using this technology. However, fear of lawsuits has prevented the wide spread use of this technology, which is not fully controllable. In some cases, the drought was ended with severe flooding and property damage.

In 1977, an international Convention was ratified by the UN General Assembly which banned 'military or other hostile use of environmental modification techniques having widespread, long-lasting or severe effects.' Here is a link to the Convention on the Prohibition of Military or Any Other Hostile Use of Environmental Modification Techniques:

"Weather warfare" is the ultimate WMD with the potential of destabilizing an enemy's ecosystem, destroying its agriculture, disabling communications networks. In other words, ENMOD techniques can undermine an entire national economy, impoverish millions of people and "kill a nation" without the deployment of troops and military hardware. Which is why its use is banned by the U.N. An informative article is here: http://www.globalresearch.ca/the-ultimate-weapon-of-mass-destruction-owning-the-weather-for-military-use-2/5306386

The physics of hurricanes and tornados is poorly understood by meteorologists. Updrafts of air over warm water create low pressure areas. Coriolis forces cause air masses to rotate around these low-pressure areas in a counter-clockwise fashion in the northern hemisphere. As air masses flow toward the center of these low-pressure areas, conservation of angular momentum causes the air to rotate faster creating an eye of the hurricane. This much is known.

But there are many unanswered questions with this simplistic model. For example, what happens to the heated air after it rises to the stratosphere? There, it should create a high-pressure area with the air rotating in the opposite direction. To my knowledge no such phenomena has been observed.

Persons unlucky enough to have seen tornados pass directly overhead have observed circular lighting on the inside funnel. Victor Shauberger did a lot of research with vortex motion and observed electrical, diamagnetic and levitation effects occurring within high speed vortexes. This was developed by him into flying saucer technology. Videos of tornados have shown heavy objects like busses and cows rotating in the air around tornados a number of times. Were they actually being levitated? Also, researchers have observed hyper dimensional, Philadelphia Experiment type effects of tornados, like straw becoming embedded inside of glass.

Hurricanes usually start with nimbus clouds that can reach as high as 100,000 feet elevations. This creates an electrical path between the positively charged ionosphere and the negatively charged earth, as explained by scientist, James McCanney. Both huge hurricanes and the smaller tornados have an electrical current nature to them and some experimentors have found ways to ground out tornados by shooting rockets trailing grounded wires behind them. James McCanney gave the Navy plans for dissipating hurricanes by a system of bouys that, on command, would release balloons trailing wires raising to thousands of feet which could ground out and dissipate the hurricane before reaching shore. If these systems were put into regular use, billions of dollars of property damage could be averted.

HAARP related technologies can also trigger earthquakes. To understand how this works, one needs to read the works of Thomas Bearden, who has written several books on the subject, like *Energy From the Vacuum* and *Fer De Lance*.

Scalar Electromagnetic waves can be created by combining two opposing (180 Degrees out of phase) transverse electromagnetic waves. Conventional Electromagnetic theory says that the two waves would cancel each other out. Bearden, however states that this would create a Scalar Electromagnetic wave which oscillates in the time domain.

These Scalar Waves can penetrate conducting mediums and can be directed into the Earth. If two or more Scalar Wave transmitters are directed at a target, intersection zones are created.

Standard transverse electromagnetic (TEM) wave energy is created where the scalar waves interfere and "energy bottles" in the zone of intersection can cause tremendous amounts of TEM energy to be stored or removed from these "energy bottles". According to Bearden, the TEM energy is trapped in the intersection zone and is not radiated away. And, this energy can be used to trigger earthquakes in fault zones already under stress.

The HAARP project was officially disbanded in 2014. However, miniaturized SBX HAARP X-band arrays have been developed into spherical microwave antennas which sit atop huge floating barges which can be towed anywhere there is an ocean: http://en.wikipedia.org/wiki/Sea-based_X-band_Radar

Also, *Aviation Week* reported in 2008 that an airborne version of HAARP had been developed which is towed 'behind a helicopter.' So, the project is probably still secretly ongoing by the military.

My research revealed many other inventors of free energy machines which also was suppressed by various means, including murder. This is why we aren't driving free energy cars and have to pay utility bills. Some very powerful companies don't want free energy machines on the market and will stop at nothing to insure they will not be.

While scientists like Nikola Tesla and Thomas Brown are virtually ignored in industry and the Universities, the military has secretly hired and used the science developed by these two men. Both men were hired to work on Project Rainbow (more popularly known as the Philadelphia Experiment), which sought to make ships invisible to enemy submarines during World War II.

Although the Navy denies that the Philadelphia Experiment was done, a large body of evidence and eyewitnesses shows that it was indeed done by the Navy. The outcome of the experiment which made the Destroyer Escort Eldrige disappear, had disastrous effects on the crew.

Most of the survivors were committed to mental hospitals. The Navy discontinued the experiment. This human disaster likely was the reason for Thomas Brown to also suffer a mental breakdown after the experiment and was laid off for a while to recuperate. Tesla had warned about the danger to the crew and wanted more time to make things safe, but was told that it was a war time emergency and that the experiment would go on anyway. At that point, Tesla resigned from the project.

However, some amazing new physical principals were experimentally proven by this experiment. And after the war, a secret Research and Development (R&D) program called Project Phoenix was headed by John von Nuemann to work out the bugs and develop teleportation technology that safely could be used by humans.

Furthermore, Admiral Rico Botta, who was partially involved with the Philadelphia Experiment, later worked near the Philadelphia Naval yard to assist the development of an experimental drive system for the future secret Navy space fleet, as revealed by William Tompkins and my book, *Inside the Secret Space Programs.*

Andrew Besagio would later claim he was teleported in both space and time – first in Project Pegasus in the 1970s with technology developed by Nikola Tesla and later in the Mars Jump Room, using technology given by the Greys, in the 1980s. This is an excellent example of secret exotic technology.

Thomas Brown would later send a proposal called Project Winterhaven to the U.S. military in which he proposed using his discoveries to create electrogravitic Mach 3 fighter crafts.

Although, the military declined Brown's proposal, they secretly developed it in black projects leading to, among other things, the B-2 stealth bomber, as revealed in the book, *Secrets of Antigravity Propulsion* by Paul A. LaViolette. This book is recommended reading for those wishing to have a better understanding of electro-gravity and its extended applications.

In addition to these technologies, there is the technology used in the secret space programs revealed by men that worked and fought in these programs, like Randy Cramer and Corey Goode.

Randy Cramer claims that he as a child he was secretly trained to be a "super soldier", along with 300 other children, 20% of which were girls, in a program called Operation Moon Shadow. Operation Moon Shadow was a USMC special section program operated jointly with an extraterrestrial group known as the Bronze Ones.

The secrecy was maintained using time travel technology. He would be abducted at night from his bed, using teleportation technology in a program known as MILABS (for military abductions). He would do his training and then he would be sent back in time, to the time he was abducted and returned to his bed. So, no one would realize that he was even gone. Also, it would seem like he only dreamed about his training in his sleep. If he told his parents about his training they would only think that he had an active imagination.

Later in 1987, he was officially transferred into a special division of the Marine Corps at age 17 and after completing his training, was sent to Lunar Operations Command, a base on the far side of the Moon. At this base, he had to sign a special contract. The contract stated that he would serve 20 years in the Earth Defense Force. At the end of his 20-year term of service, he would be age regressed 20 years, his memory of his service would be wiped out and he would be sent 20 years back in time to start a new life. This tour of duty was known as "20 and back".

Apparently, the Earth Defense Force is a UN "Unacknowledged Special Access Program". This defense force recruits people from many military services globally. After signing this contract, Randy Cramer and others were sent to Mars to serve in the Mars Defense Force. This force would defend the colonies operated there by the Mars Colony Corporation. The enemy there were primarily Reptilians and Mantids.

He described the *Archer Series, Full Body Power Assisted Armor and Environment Suit*. This suit not only provided armor but also was light weight, and had a self-powered exoskeleton that enhanced the soldier's strength and assisted breathing in the thin Martian air.

The helmet for this suit had a heads-up display with multi

spectral inputs, infra-red, night vision and sonic. This display could detect incoming projectiles from over one thousand yards distance and light them up on the visor. This allowed an alert soldier with good reflexes to actually dodge incoming bullets. This secret technology is far in advance of what the marines are equipped with here on Earth.

Randy Cramer was blown up in battle many times. He explains that for example, if a soldier had his arm blown off above the elbow, the highly trained medics would use what Randy Cramer called "a high definition holographic cellular regenerative technology machine", to totally regrow a new arm in 5 hours in the hospital.

Amazingly, the soldier would be ready for battle the next day. His and his comrade's lives were saved many times with this advanced medical technology. Imagine how useful this type of technology would be in hospitals here on Earth!

Corey Goode discusses teleportation technology on the Solar Warden research ship, *Arnold Sommerfeld,* which could teleport him without a receiver on the other end. And, this technology could even teleport him to underground bases through solid rock and back on board after his mission was completed.

This ship used gravity plates to create a 1 G artificial gravitational field while out in space. Also, there were replicators on board which they termed "printers". He used the printer to create his meals. He could select his meal by pressing a selection button on an electronic screen. Roast Beef was Corey Goode's favorite.

Then, there were the smart glass pads which stored the true history of the Earth in great detail, which Corey Goode would study during the many boring hours of his twenty-year tour of duty, totally cut off from contact with Earth for security reasons.

This gives you some idea of the secret exotic technology that has been hidden from the people of the Earth. And, I am sure that there is plenty more that I don't know about.

Think about how this technology could be used for the benefit of the Earth. Pollution free energy would quickly clean up the threatened ecology of the planet. Health care would be revolutionized. Transportation would be quick and efficient and non-polluting. Hunger would be a thing of the past. The list is only limited by your imagination.

These are the reasons for pushing for a full disclosure of these secret exotic technologies that are now being withheld from the people of our planet.

Conclusions

Although the nineteenth century instituted the industrial revolution. It didn't really get going until the twentieth century. That is when industrial manufacturing started to overtake agriculture as a large source of wealth creation.

Large corporations, monopolies and cartels were forming, like General Motors, Standard Oil and I.G. Farben. Also, a monopoly in the form of central banks were exercising excessive amounts of financial power.

And, these corporations started to have more actual power than governments – particularly democratic governments, which were always changing leadership based on the evanescent will of the people.

As a way of example of this corporate power over government, the following account may prove enlightening. In 1942, after the U.S. entered the war against Nazi Germany, it was discovered that Standard Oil of New Jersey Corporation (the later EXXON) was still refueling German ships in Aruba in the Caribbean.

A meeting was convened to discuss how Standard Oil should

be punished for their treason in giving aid and support to the enemy. After a lengthy discussion on the matter, it was decided that the U.S. military machine was so dependent on Standard Oil that punishment could jeopardize the war effort. So, no punishment was forthcoming.

Another example was" the too big to jail' industrialists who plotted a military coup against the Roosevelt Administration. Clearly, these people were virtually more powerful than the U.S. government and were above U.S. law. We also saw how the corporate news media used "spin control" to calm the public about this act of unconstitutional crime. This "spin control" has been a continuing feature of the corporate news media corporations.

One definition of fascism is a form of radical authoritarian nationalism, characterized by dictatorial power, forcible suppression of opposition, and control of industry and commerce, that came to prominence in early 20th-century Europe.

The interesting thing about this fascism was that Mussolini and Hitler both encouraged privatization of formerly government owned institutions. This is much like the Republican strategies to privatize government functions in the present-day U.S., such as prisons, highways, the postal service and even public education!

Many U.S. industrialists seemed to have a love affair with this fascism - not only before World War II - but also after the war! Of course, after the war, they had to be more circumspect about their true feelings due to the public's horror of Nazi atrocities.

Many of these U.S. companies had subsidiaries in Germany during the war which used slave labor to increase their profits.

In the present, when they can get away with it, they still use slave labor, as the ICC does in their secret Mars colonies. And, as many prison laborers in the U.S. are used. And the offshoring of U.S. jobs to other countries that pay slave wages is an ongoing phenomenon.

Now to be fair, corporations do provide us with many useful things, like cars, refrigerators, electronic devices and so on. And obviously, governments are too inefficient to manage most corporate operations profitably, as the former communist Russia discovered. And, not all corporate CEOs are so worshipful of Mammon, the money God, that they sell their souls for profit at all human costs.

The main problem is with those companies that want profitability regardless of the human and planetary ecological costs of their operations. And many of these companies are the same ones that were involved with unconscionable activities during World War II and afterwards.

But, the social human issues will be unavoidable when robots start doing most of the labor previously done by humans in the future. Yes, robots that work 24 hours a day, seven days a week, without pay, sick leave of vacation time off, will dramatically increase the productivity of factories while relieving humans of boring repetitive assembly line work.

The end effect of this robotic revolution will be to change the whole money paradigm. What good would all this productivity be if there was no market for the products? There would be no market if all the laid off former workers of these now robotic factories had no money to purchase these products. So, the economy depends as much on people with money to spend as with actual production. This unescapable fact will create a different way of looking at the capitalistic theory of government.

The communistic theory has proven to be impractical because it lacks sufficient human incentive for productivity. The capitalistic theory will also prove impractical because it eventually concentrates all the wealth to a few - at the expense of the many. Those who play the game of Monopoly know that when one person has all the money - the game is over.

This will also get even worse with the robotic revolution because those with the capital can purchase the robots needed to upgrade these factories and those without capital will have to keep the human workers. Competition will soon place the non-robotic factories out of business and there will be even less people with money to purchase the robotic factory production.

And that will be the time that the capitalistic fallacy will become all too evident and without some form of socialistic government intervention, the whole economy will collapse on itself.

So, with fascism governments and corporations work together in a dictatorial way with little regard for human rights. And with communism, workers are supposed to have some control over the state-owned corporations. But, the personal incentives for production are missing. And again, when too much power is given the state, as with Stalin's Russia, human rights again suffer.

Socialism seems to be the middle ground between communism and capitalism. Europe has worked with socialism quite successfully. To be sure, there is high taxation in socialism. However, there are also benefits, like free health care, longer vacations, less crime, less homelessness and less other problems facing the U.S. capitalistic system.

Looking ahead, Europe and the State of Hawaii, already have floated the concept of a Universal Basic Income in their legislatures. But, the idea seems ahead of its time as it was voted down.

But, The Universal Basic Income is a concept that would place sufficient income into an individual's bank account to insure basic survival needs, like food, clothing and shelter. For luxury items, the individual would have to earn his or her own extra money. This concept would also help to prevent an economic collapse when robots displace most human jobs. I predict this idea will have more traction in the future, as more robots and self-driving trucks and cars are put to use.

Another issue is full disclosure of all the government's and breakaway civilization corporations' secret programs. A breakaway civilization corporation is defined as an entity that is independent of the existing civilization with its laws and economic limitations and whose existence, because of high secrecy, is virtually unknown to the existing civilization.

The Interplanetary Corporate Conglomerate (ICC) is not traded on any stock exchange nor does it pay any taxes to any Earth based government. Nor is its existence known to most.

Yet, the ICC has many underground facilities in this planet and facilities throughout our solar system, and does a considerable amount of trade with many extraterrestrial civilizations, as revealed by Corey Goode in his many presentations, Tony Rodriguez in his interviews with Michael Salla, and others.

The ICC has the ideal setup (to the mind of a CEO) on Mars and other interplanetary locations, because their company and its subsidiaries, like the Mars Colony Corporation, use slave labor to maximize their profits. For example, on the Mars colony, the workers are virtual prisoners that can never return to Earth. The Mars Defense Force that protects these Mars colonies is comprised of fighters that have signed "20 and back" tours of duty and are more like 20-year bond servants/soldiers than complete slaves because they eventually regain their freedom.

According to Toney Rodrigues, the large asteroid, Ceres, is being mined. The mining is done by excavating large underground caves in the ore deposits. After the mining is completed, the excavated caves are turned into underground cities, complete with living quarters, factories, entertainment centers, stores and space ports. From these space ports trade is carried on with other star systems, using extremely advanced technology.

However, with all the advanced technology being used there, the form of civilization there is very backward and regressive. Human rights there, are practically non-existent.

According to Rodrigues, the place is run just like Nazi Germany was. When he first arrived on Ceres, he and his group were informed that they would have to obey orders and if not, they would be shot on sight. The Nazi swastika was displayed everywhere. They also had to give the stiff-arm salute and say Hiel (followed by the name of whoever was the head CEO of the Ceres operation – which would change every so many years) when ever given orders by a superior.

Also, Rodrigues had to wear a shocker collar so that if a superior felt that Rodrigues required punishment, he would be given a severe electric shock. Rodrigues and others in his group had to humbly submit to these conditions because there was no escape from Ceres. The men knew that they just had to put up with their conditions until their 20-year tour of duty was completed.

So, we see the Nazi like nature of these corporations. As the Nazis used a lot of slave labor during World War II, the ICC also uses slave labor in the present. Perhaps, this could change with the wider use of robotic manufacturing. But, perhaps not.

What would the ICC do with all their slaves when they are no longer required because robots can do their work more efficiently - free them and send them back to Earth? But then, these freed slaves might sue for damages. The least expensive option would be to simply exterminate them – perhaps in gas chambers. CEOs of corporations often choose the least expensive, most efficient option.

A more likely and positive scenario would be the informed and angered public demanding that the ICC be punished for using slave labor and having the entire use of slave labor outlawed, off the Earth as well as on the Earth. And, that is one reason people like me bother to research and write books on these highly secret and criminal operations – so that human justice and fairness will someday prevail!

What can we as individuals do to make our world a better place, to live while putting a stop to the senseless, greed driven destruction of our planet via war, pollution and a proliferation of products that are unhealthy for us and our children, like GMO food and fluoridated tooth paste.?

One thing we all can do is to make an effort to raise our level of awareness of the causes of the problems that are besetting our society and planet. This means raising above our perceived immediate survival needs and worries to a more inclusive outlook of our world. The world is a spherical whole – not separated, man-made countries. It is a life supporting environment which must be protected and preserved so that it can continue to support life.

This issue is more important than corporate profits, as mankind is coming to realize. Once our basic needs are met, quality of life becomes more important than quantity of money in a bank account. The quantity of our possessions has absolutely nothing to do with the quantity of our happiness, as I realized at an early age.

This involves becoming more informed and less manipulated by the mass media that is always trying to sell us something, whether it is a product, religion or a political point of view. The mass media are corporations which, like most corporations, usually are more interested in their own bottom line than our best interests. May I suggest reading some good books in your area of interest and watch less T.V.?

There are many civilizations in our Galaxy and Universe, some of these civilizations have already made contact with our civilization both through individual contactees like George Adamski, William Meir and Howard Menger, and through government leaders.

Unfortunately, the governments have gone to great lengths to hide these Extraterrestrial contacts from the public – the primary reason for public non-awareness of this fact.

However, much can be learned about these extraterrestrials by reading books written by the individual contactees. These books can easily be found by placing the contactee name into an Amazon.com, or another online book seller search engine.

These types of books usually describe the positive extraterrestrials that promote peace and universal brotherhood. However, there also exist negative extraterrestrials that promote selfish, greed orientated philosophies that inevitably lead to war and destruction. And unfortunately, the negative E.T.s are the ones that Earth governments made secret treaties with.

A universal law of the universe is that more advanced civilizations are not to interfere with less advanced civilizations. This interference could prevent the evolving civilization from learning the most favorable way of doing things, much as over-indulging or protective parents might prevent their children from learning wisdom by shielding them from painful life experiences. The Star Trek series presented this law as the "Prime Directive".

This Prime Directive can however have exceptions. One exception is in the case of the possibility of a civilization destroying their own planet. The positive extraterrestrials had approached Earth government leaders and tried to persuade them to give up nuclear weapons.

The Earth government leaders that possessed nuclear weapons thought it was a ploy to disarm them and thought that nuclear weapons might be the only weapon they could use against the extraterrestrials superior technology in case they attacked Earth. These Earth based governments requested some of the advanced technology of the positive extraterrestrials. They declined, stating that Earth humans already possessed technology that they lacked the wisdom to use correctly and warned that the Earth was on a path to destruction. So, these Earth government leaders decided not to do business with the positive extraterrestrials.

Another exception to the Prime Directive is in the case when the lesser evolved civilization actually asks for involvement with the more evolved civilization. According to the wisdom level of the more developed civilization, they may refuse or accept the lesser evolved civilization's request.

In this case, the negative Draco Reptilians, acting through their allies, the Orion Greys, decided to make deals with some of the Earth based governments, which included Nazi Germany, and the United States. These deals usually traded technology for the rights to do genetic experimentation on humans.

These Greys did not request the giving up of nuclear weapons after the nuclear age. So, these were the extraterrestrials that Earth Governments decided to do business with, rather than the positive extraterrestrials. Then, these governments went to great lengths to keep it all secret.

Antonio Urzi is one of the most recent UFO contactees to be in the public. Urzi has captured many UFOs on his video camera. He has almost 100 films. He has seen spheres as well as the typical saucer shape UFO. Many of his films tend to be higher quality and closer than most UFO films.

He said they often come to him and sometimes his intuition is guided by the space people (like Adamski) to see their UFOs. Urzi feels a feeling of peace coming from the UFOs. Many times, the UFOs show up to him as if they are posing for him to film them as with as they did with Billy Meir.

The Urzi UFO films have been proven to be real by Jim Dilettoso, a computer expert who has worked at The Village Labs in Arizona and throughly analyzed the documentary *The Urzi Case A UFO Mystery In the Skies of Italy* by Pier Giorgio Caria.

The positive extraterrestrials contacting individual Earth contactees are nominally from an alliance called the Galactic

Federation of Light. The real federation will never give out any high level-information about their federation because we are not members of their organization.

The Galactic Federation of Light are the space friend organization that contacted George Adamski, Howard Menger, Buck Nelson, Daniel Fry and was involved with the Italian Friendship Group.

In the book *50 Years of Amacizia and Mass Contacts* by Stefano Breccia, the space friends also call their organization the Confederation. This book documents 50 years of UFO contacts with humans. The UFOs mentioned in this book are quite likely the same ones filmed by Antonio Urzi.

Howard Menger's space friends are believed to the same group from the same confederation due to the similarity of the ships and the way they taught.

The teachings of our space friends are quite like the teachings of Jesus Christ and are relayed to individual contactees that our space friends consider to be sufficiently spiritually evolved to relay the teachings, without ego problems, to the rest of mankind. When these teachings are put into every day practice by the majority, Earth will become a true paradise.

Many will state that there is no scientific evidence of spirit and that spirituality is in conflict with scientific principles.

Spirituality is based on direct experiences of ordinarily invisible numinous dimensions of reality, which become available in holotropic states of consciousness. It does not require a special place or officially appointed persons mediating contact with the divine. The mystics do not need churches or temples. The context in which they experience the sacred dimensions of reality, including their own divinity, is provided by their bodies and nature. And instead of officiating priests, they need a supportive group of fellow seekers or the guidance of a teacher who is more advanced on the inner journey than they are themselves.

Organized religions tend to create hierarchical systems focusing on the pursuit of power, control, politics, money, possessions, and other worldly concerns. Under these circumstances, religious hierarchy as a rule dislikes and discourages direct spiritual experiences in its members, because they foster independence and cannot be effectively controlled. When this is the case, genuine spiritual life continues only in the mystical branches, monastic orders, and ecstatic sects of the religions involved.

A deep mystical experience tends to dissolve the boundaries between religions and reveals deep connections between them, while dogmatism of organized religions tends to emphasize differences between various creeds and engenders antagonism and hostility, rather than true peace and love practiced and experienced by truly spiritual persons.

There is no doubt that the dogmas of organized religions are generally in fundamental conflict with science, whether this science uses the mechanistic-materialistic model or is anchored in the emerging paradigm. However, the situation is very different in regard to authentic mysticism based on spiritual experiences.

The great mystical traditions have amassed extensive knowledge about human consciousness and about the spiritual realms in a way that is similar to the method that scientists use in acquiring knowledge about the material world. It involves a methodology for inducing transpersonal experiences, systematic collection of data, and intersubjective validation. Spiritual experiences, like any other aspect of reality, can be subjected to careful open-minded research and studied scientifically.

Scientifically conducted consciousness research has brought convincing evidence for the objective existence of the imaginal realm and has thus validated the main metaphysical assumptions of the mystical world view, of the Eastern spiritual philosophies, and even certain beliefs of native cultures. The U.S. and Soviet military's use of remote viewing programs use some of these Eastern meditational practices.

The conflict between religion and science reflects a fundamental misunderstanding of both. There cannot be a conflict between science and religion if both these fields are properly understood and practiced.

If there seems to be a conflict, we are likely dealing with bogus science and/or bogus religion. The apparent incompatibility is because either side seriously misunderstands the other's position and very likely represents also a false version of its own discipline.

Also, some scientists are actually practicing a religion or a belief system rather than true science. They automatically disregard experimental evidence that goes against their scientific beliefs. True scientists understand that the outcome of a properly organized experiment is the true judge of reality.

In any case, each one of us can make a positive contribution to preserving our planet by the way we live our lives – regardless of what our government is doing right or wrong. The way is quite simple as many spiritual teachers have taught.

We are truly brothers and sisters, in that we all are sons and daughters of God, the creator of all that is. Any man-made religion which states that their religion is superior to another's or that tries to create separation between different groups of mankind is presenting a false doctrine. This concept also applies to politics. Any politician that creates division by religion, race or financial standing is also putting out a false doctrine. These false doctrines should be avoided like the plague because that is what they are – a mental and spiritual plague. By understanding that simple fact, we can avoid many pitfalls in our understanding of the world.

The golden rule is to treat another as you would want them to treat you. Every truly, practicing Christian knows this. The same applies to all other religions.

Practice the golden rule in your everyday life - no matter what religion you practice or believe in or even if you are an atheist. Simple logic will tell you that if everyone did this, it would immediately solve many of societies' problems and make the world a better place to live for everyone.

Another good practice is to, in your mind, place yourself in someone else's circumstance. That would help you to have a better understanding of where they are coming from and have and a greater compassion for their situation.

Don't be judgmental of others. Have you ever noticed how people making judgements of other people often suffer from the same "faults" they see in others?

As true spiritual teachers have said, "As you judge others, you are judging yourself." In other words, a fault you see in another probably also exists within yourself.

If you are tempted to make a judgement on another - also attempt to see where you might have a similar fault and try to do the work on yourself to overcome that fault. This is much harder to accomplish than to be said but, is truly a worthy goal.

There is another aspect of judgement. It is difficult to truly understand why a person does what they do, unless we ourselves have experienced the same or similar things that the other person has.

There is a saying that "birds of a feather flock together". This saying applies to people. People with similar interests, desires and outlook tend to band together. They put out an "energy" that attracts others with a similar "energy". So, people with criminal intent are attracted to others with criminal intent. People with a helpful intent are attracted to others with a helpful intent. Politicians are attracted to other politicians and so on. The point is to become more aware of the type of energy you are putting out because that is the type of energy you will attract to yourself!

Forgiveness helps to eliminate karma and is a cleansing for your own heart or emotional body. People who wish for justice or revenge are unknowingly keeping themselves on the wheel of karma. Simple logic shows that societies that demand revenge will have more violence, which will keep escalating, than societies that practice forgiveness.

According to Jesus Christ, the primary law is that of Love. "Love thy God with all thy heart and understanding and love thy neighbor as thy self." If you can practice this law of love, all other laws will be fulfilled. This love should include a respect for all the Creator's living creations and our planet itself.

Love is a more powerful force than fear. Let your motivations be ones of love rather than fear and your life will be infinitely more fulfilling, while making the world a better place to live. Our space brothers also say the same thing.

God is not a man or a crucifix. God is the invisible intelligent energy that is in all people and all things and is the source of all consciousness and life. All advanced civilizations in the universe believe in God but they do not have organized religion because their advanced science, which includes consciousness and spirit, already incorporates the concept of God. Organized religion, in the last analysis, is really a political process to control people.

These teachings, when put into practice in our everyday lives form a system of morality that keeps us from being selfish, criminal, barbarians and allows civilization to flourish in peace and harmony. For some, morality may seem boring or unnecessary. But, morality is truly the cornerstone for a peaceful society.

We can also use prayer. The whole universe exists within God. Ask God for assistance in your life and to help all people all over the world have morals and forgiveness. This will really have an impact as everything is connected by God's energy.

We are just temporary visitors of the Earth – spirits living in human bodies. We cannot take our amassed material possessions with us into the next world after we transition from our human body in the process we call death. We can however take our, talents, experience and wisdom there.

So, "what is the purpose of life?" you might ask. The purpose of our life in this material world is to learn from first-hand experience. The learning on the material plane, which is not our true home, occurs much faster than on the higher more ethereal planes of existence. This is because of pain and the fixed nature of matter which forces us to focus our attention and deal with the consequences of our decisions. This is our spiritual testing ground.

God is infinite. The whole universe exists within God. God is infinite. So, the learning about the universe is also infinite. The learning never ends. Our soul (our awareness) is a part of God. So, our soul is also infinite. When we have mastered the material plane, we can go on to higher planes of experience and learn to master those planes also.

End

Other Books by This Author

The Adventurer https://www.amazon.com/Adventurer-
Autobiography-Herbert-Grove-Dorsey-ebook/dp/
B00547RVQK/

The Secret History of the New World Order https://www.
amazon.com/Secret-History-New-World-Order-ebook/dp/
B00LAECCJM/

Secret Science and the Secret Space Program https://www.
amazon.com/secret-science-space-program-ebook/dp/
b00pss3re0

CIA: Crime Incorporated of America https://www.ama-
zon.com/CIA-Incorporated-Herbert-Dorsey-III-ebook/dp/
B013RJEJ20/

The Covert Colonization of Our Solar System https://www.
amazon.com/Covert-Colonization-Our-Solar-System-ebook/
dp/B01C0BLUUA/

Inside the Secret Space Programs https://www.amazon.
com/Inside-Secret-Programs-Herbert-Dorsey-ebook/dp/
B06X95DM13/

Bibliography

1. *Pawns in the Game* by Admiral Guy Carr

2. *Terrorism and the Illuminati* by David Livingstone

3. *Secret Science and the Secret Space Program* by Herbert G. Dorsey III

4. *The Vatican's Holocaust* by Avro Manhattan

5. *The Secret History of the New World Order* by Herbert G. Dorsey III

6. *Wall Street and Hitler's Rise to Power* by Anthony Sutton

7. *Selected by Extraterrestrials* by William Tompkins

8. http://www.ufocrashbook.com/eisenhower.html

9. *Inside the Secret Space Programs* by Herbert G. Dorsey III

10. http://worldnewsdailyreport.com/monsanto-creates-first-genetically-modified-strain-of-marijuana/

11. *Vietnam Why Did We Go?* by Avro Manhattan

12. *Selected by Extraterrestrials* by William Tompkins

13. https://www.amazon.com/ALTERNATIVE-Sci-Fi-Classic-Republished-Material-ebook/dp/B0182YIKNE

14. http://exopolitics.org/20-years-a-slave-in-secret-space-pro-grams-abduction-programming/

15. https://borderlandsciences.org/cart/flying-saucers/#EdwardsSaucers

16. *Secret Journey to Planet Serpo* by Len Kasten

17. http://www.serpo.org

18. *Lost Science* by Gerry Vassilatos

19. *Secrets of Cold War Technology: Project Haarp and Beyond* by Gerry Vassilatos

CPSIA information can be obtained
at www.ICGtesting.com
Printed in the USA
LVHW112332100719
623749LV00001B/75/P

9 781478 793991